THE MACROFUNGI OF ANDEAN PERU Part 1

AN OVERVIEW OF MACROFUNGI IN PART 1

AND ASSIGNED (P) NUMBERS

FAMILY AGARICACEAE

Agaricus

P1 P2 P3 P4 P5 P6
P7 P8 P9 P10 P11 P12
P13 P14 P15 P16 P17 P18
P19 P20 P21 P22 P23

Coprinus　　Hymenagaricus　　Melanophyllum　　Micropsalliota

P24 P25 P26 P27 P28 P29

Gastroid Agarics

Agaricus　　Bovista

P30 P31 P32 P33 P34F P35

Calvatia

Disciseda   Lycoperdon

Podaxis   Tulostoma

Cystoderma   Chamaemyces   Cystolepiota

THE MACROFUNGI OF ANDEAN PERU Part 1

Lepiota

P73 P74 P75 P76 P77 P78

P79 P80 P81 P82 P83 P84

P85 P86

Pseudobaeospora

P87

Chlorophyllum

Leucoagaricus

P88 P89 P90 P91 P92 P93

P94a P94b P94c P95 P96 P97

Leucocoprinus

Macrolepiota

P98 P99 P100 P101

FAMILY AMANITACEAE

Catatrama

Limacella

Saproamanita

P102 P103 P104

FAMILY BOLBITIACEAE

Bolbitius

Conocybe

Galerella

Panaeolina

Pholiotina

FAMILY CLAVARIACEAE

Clavaria

Clavulinopsis

Indetermined choral fungi

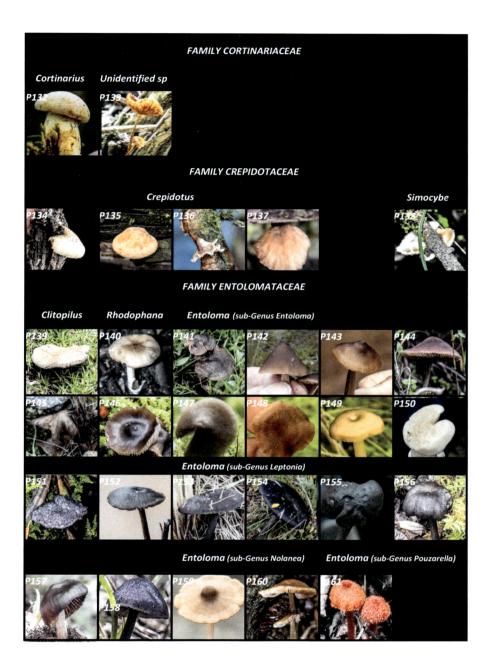

FAMILY CORTINARIACEAE

Cortinarius   Unidentified sp

P132   P133

FAMILY CREPIDOTACEAE

Crepidotus   Simocybe

P134   P135   P136   P137   P138

FAMILY ENTOLOMATACEAE

Clitopilus   Rhodophana   Entoloma (sub-Genus Entoloma)

P139   P140   P141   P142   P143   P144

P145   P146   P147   P148   P149   P150

Entoloma (sub-Genus Leptonia)

P151   P152   P153   P154   P155   P156

Entoloma (sub-Genus Nolanea)   Entoloma (sub-Genus Pouzarella)

P157   158   P159   P160   161

FAMILY HYDNANGIACEAE

Laccaria

P164   P165   P166

FAMILY HYGROPHORACEAE

Chromosera   Gliophorus   Humidicutis

P167   P168   P169   P170   P171

Neohygrophorus   Hygrocybe

P172   P173   P174   P175   P176   P177

P178   P179   P180   P181   P182   P183

P184   P185   P186   P187   P188   P189

P190   P191   P192   P193   P194   P195

Hygrophorus   Arrhenia   Cora

P196   P197   P198   P199

Pseudoamillara

P200

Lichenomphalia

P201

Cuphophyllus

P202 P203

Cuphophyllus (cont)

204 P205 P206 P207 P208

FAMILY HYMENOGASTRACEAE

Galerina

P209 P210

Gymnopilus

P162 P163 P213

Psilocybe

P211 P212

FAMILY INOCYBACEAE

Inocybe

P214 P215

FAMILY LYOPHYLLACEAE

Calocybe

Gerhardtia

P216 P217

PREFACE TO PART 1

This compendium in English is the product of ten years of traveling, collecting, analyzing and documenting the findings of the first collection of mushrooms and macrofungi found in an area spanning the length of the Peruvian Andes from Puno in the South to Tumbes in the North. As such it represents but a fraction of the total macrofungal diversity in this large expanse of mountains but it is a start, an initial base to build on. The compendium does not include information from other valuable smaller and localized existing collections but only the samples collected in the survey. The English version presented here is Part 1 of a three-part series that documents the individual fungi found and provides initial identifications. It is hoped it will help future investigators, educators, decisionmakers and nature lovers promote and protect the unique diversity and stunning and critical natural ecosystems of Andean Peru. The English compendium is based on an-as-yet unpublished 800-page Spanish draft developed as the basis for a report to the Peruvian Government in 2021. It is planned to publish the Spanish version later.

The collection reported here began following initial research to elucidate the sparsely documented status of traditional mushroom consumption in the Peruvian Andes where malnutrition is elevated. Mushrooms and their consumption could then be compared to other mountainous regions of the world where mushrooms are important food (Boa 2004). It turns out that there exists an extensive and ancient tradition of mushroom consumption in Peru that is rapidly disappearing. However, the mushrooms consumed are taxonomically very different to those known in the northern hemisphere (Holgado 2010, Trutmann and Luque 2012). Unfortunately, a lack of scientific documentation of native macrofungal diversity of the Andes has severely slowed the ethnomycological efforts linked to these ethnomycological studies. This underlined the fundamental need to make available more complete information on fungal diversity in the Peruvian Andes to facilitate mycological research on important human and environmental health issues.

With this problem in mind, we connected with national mycologists and ecologists in Peru. The first was Magdalena Pavlich Herrera, a renowned fungal taxonomist based at the Universidad Peruana Cayetano Heredia (UPCH) in Lima, who had already conducted an important taxonomic study of Andean and Selva regions in the 1970s (Pavlich Herrera, 1976). The second was María Encarnacion Holgado Rojas at the Universidad Nacional San Antonio Abad de Cusco (UNSAAC), who had begun to document the ecological and ethnomycological diversity of mushrooms in Cusco (Holgado Rojas et al., 2007). In the north we teamed up with Mario López Mesones in Chiclayo a well-known ecologist who had documented the ecological importance of northern Andean highland forests (Llatas-Quiroz and López-Mesones 2005). Together with them, and students in Cusco, Amarilda Luque and I began an effort to collect baseline information on the macrofungal diversity and distribution in the Peruvian highlands from the Bolivian to Ecuadorian boarders. The macrofungi in the collection are described morphologically, but for various reasons have not yet been subjected to molecular characterization. Nevertheless, it is hoped the compendium serve as a useful reference and source to access collection material in the coming years to help unravel more of the mycological and environmental riddles and stimulate more appreciation of the rich fungal diversity of the magical Andean region of Peru.

ACKNOWLEDGEMENTS

During the course of the collection and identification efforts there have been many individuals who made important contributions to its completion, all of whom I would like to acknowledge and thank.

First, as principal author I would like to acknowledge and thank the coauthors of this compendium, in particular, Amarilda Luque Luque M.Ag.Sc., of the *Centro Para el Procesamiento de Datos Estadisticos* (CPROESTAD) and *Global Mountain Action* (GMA), who accompanied me in most of collections, in the most inhospitable as well as breath taking places and who worked with me often in dingy hotels until late in the night and before sunrise to process fresh collections during the years of collection and patiently stuck with me throughout the long periods of analyses, identification and finally documentation of the work. Grateful thanks also go to Dr. Magdalena Pavlich Herrera, the fungal taxonomist and professor of microbiology at the *Universidad Peruana Cayetano Heredia* (UPCH), who opened her laboratory for the work at the beginning, provided taxonomic guidance and access to important taxonomic literature, as well as a home in the herbarium of the UPCH for the collection samples. Above all, she steadfastly supported the work over the years and has been not only a colleague but a dear friend, who in the end provided important feedback on how to better organize the document from a taxonomic perspective. Similarly, I am grateful to Dr. María Encarnacion Holgado Rojas, the mycologist and professor and friend at the *Universidad Nacional San Antonio Abad de Cusco* (UNSAAC), who from the start provided generous collaboration and support that enabled the collections in otherwise inaccessible places around Cusco, provided students and who opened her laboratory to us. My thanks also to Albino Quispe Pelaez (now completing his PhD) and Milton Kallañaupa Auccapuma, both UNSAAC alumni students, who provided invaluable help finding fungi using their local knowledge and connections that significantly helped the collection around Cusco. And, last but not least, I am grateful to Maestro Mario López Mesones, professor at the *Universidad Nacional Pedro Ruíz Gallo* (UNPG), Chiclayo in Lambayeque, a colleague and friend, who guided us to and collected with us in the disappearing relic high selva Amazonian forests in the north of the Peruvian Andes around Cañaris and Colasay. His profound knowledge and intimate trust and connections with the Cañaris people enabled collection in places we would never have been able to reach alone.

Secondly, thanks and acknowledgement to the many local people, who shared their knowledge of mushrooms and were vital to the work. A non-exhaustive list of these individuals includes, in Cusco: Sr. Daniel Benito, Sra. Wilma Monson, Sra. Justina Maywa Chacon, Sr. Juan Palladol Chaika, Sra. Felicias Valdez Mascco, Sr. Leoncio Escalante Quispe and Sra. Lorenza Sanabria; in Puno: Sra. Lydia Luque Cutipa, Sr. Hector Biamonte, Sr Porfirio Mamani Cáceres, Sr Fidel Callañaupa Quispe and Sr Rodolfo Condori; in Ancash: Sra. Victoria Villafranca, Sr. Maritza Ñupe and Sra Juanita; and in the North of Peru: Sr. Jesus Eloy Reyes Huaman, Sr.

Enrique Atoche, Luis Grippa, Sra. Doña Diana Rojas, Sr. Roberto Reyes Rinza, and Sr. José Gaspar Lucero and Sra. Marcela Sanchez.

Thirdly, although the principal author takes full responsibility for any errors in identification in the final publication, a deep gratitude to all those who generously helped in the identification of samples, who pointed out errors and helped provide best guesses for samples with incomplete information. They include: Dr. Gerardo Robledo, *Universidad Nacional De Cordoba*, Argentina; Dr. María Paz Martín, *Real Jardín Botánico de Madrid*, Spain; Dr. Teresa Iturriaga, *Cornell Plant Pathology Herbarium (CUP)*, USA; Mr. Danny Newman, specialist in South American Neotropical fungi, USA; Dr. Donald Pfister, *Harvard University Herbarium*, USA; and Dr. Scott La Grecca, *Duke University*, USA.

Fourthly, grateful thanks to a group of remarkable students of the UNSAAC Cusco for their help in collecting, for the energy and friendship. They include, but are not limited to, Karin Pérez Leguía, Mario Callalli Chancahuaña, Luis Edvardo Cano Carrasco, Karina Farfán Romero, Wilfredo Miguel Espinosa and Elías Paz Apaza among others. Thankful acknowledgements also to Ruth Lazarte, professor of Microbiology (UNSAAC), Mr. Roberto Matos, mushroom producer (Calca) and Moises Ordonez, professor at UNPRG in Lambayeque, for helping find fungi. And, of course, thanks to Aaron Luque, who although young in age accompanied us on the trips and helped collect fungi.

Last, but not least, thanks to the Peruvian Government's Forestry Service (SERFOR) for permission to collect over the years, and for the support of Dr. Stefan Flückiger, Dr. Marco Baltensweiler and Dr. Bruno Stöckli from *Global Mountain Action* (GMA), and finally to Olivia Vent and Christel Trutmann for editing and backing of the Trutmann Family to accomplish this work.

INTRODUCTION

Fungi are one of the most diverse groups of organisms on earth and constitute a significant part of terrestrial ecosystems. They form a large share of the species richness and are key players in ecosystem processes (Keizer, 1998; Seen-Irlet et al., 2007). As such, they play an essential role in the functionality of both wooded and grassland ecosystems (Stamets P., 2005). Mushrooms - or more broadly macrofungi, - defined here as fungi that produce fruiting structures visible to the naked eye, do so principally by providing mechanisms of recycling of complex biological material, through mineralization and transport of otherwise bound, unavailable compounds to plants and other microbial life. They act as a mechanism of selection to maintain a dynamic balance by eliminating weakened parts of the ecosystem - an action that is aggravated in disturbed systems (Baker, 1974) - and by forming valuable symbiotic relationships with almost all plants, commonly known as mycorrhizal associations crucial to plant growth, resistance to climate change (Usman et al., 2021) and mutual communication (Gorzelak et al., 2015). As pathogens they provide selection pressure in ecosystems weeding out weakened individuals to maintain system health. Fundamentally, without fungi ecological systems on which we depend would stop functioning. An ecosystem with a compromised fungal community has a reduced capacity to perform the above-mentioned functions and risks losing robustness and equilibrium under stress from environmental changes, including climate variability. Without fungal diversity forests and grasslands lose their productiveness and lead to less capacity of ecological systems to regulate water, carbon and oxygen, soil fertility and withstand biological stresses such as diseases.

These processes are particularly relevant to mountain systems that are amongst the most important sources of water, as well as some of the most biologically varied environments on earth (Antonelli et al., 2018). They contain some of the most fragile ecosystems and environments (Price 2006). To maintain functional mountain ecosystems in an increasingly stressed environment it is essential to maintain essential biological diversity (Mohammed 2019). Yet, to date, very little information exists about the composition, diversity and ecology of fungi of the Peruvian Andes. This makes accurate assessments of ecosystem health in mountains very tenuous as there is a lack of a base line to refer to. Without such it will be difficult to make progress in understanding the Peruvian Andean ecology, or make much progress in mycological research on issues such as climate and land use change.

Current knowledge of macrofungi in Andean Peru

The earliest known macrofungal collection in Peru was made by Fortunato Heredia between 1921 and 1932 and is housed at the Universidad Nacional San Antonio Abad de Cusco. He reported 37 Ascomycete and Basidiomycete species of macrofungi almost exclusively from the families Xylariaceae, Auriculariaceae, Cortinciaceae, Polyporaceae, and lignicolous Agaricaceae (Herrera 1941).

The most extensive published collection was made by Magdalena Pavlich (Pavlich Herrera, 1976) and is housed at the herbarium at the Museum of Natural History in Lima. Over a nine-year period she collected and identified altogether around 100 macrofungi in the departments of Loreto, San Martin, Huanuco, Amazonas, Junín, Pasco, Ayacucho, Cusco, Puno, La Liberdad, and Madre de Dios. These included nine species of Ascomycetes in eight genera, and 93 species of Basidiomycetes in 44 genera, of which 16 genera (25 species) were in the Agricales and 18 in the Aphyllophorales.

Earlier general collections were made by foreign prospectors that included some macrofungi. Ralf Singer also made rapid collections in Peru. For example, he collected and identified the traditionally consumed Q'oncha Cusqueña as *Pleurocollybia ciberia* (Singer, 1963). The status of his collections in various herbaria around the world requires clarification. Other mainly botanical collections that include macrofungi are those made by Pearce housed at the Royal Botanical Gardens (Kew), by Uhle in the Museum of Berlin, by Cook and Gilbert kept in the National Museum in Washington DC, by Killip and Smith in the New York Botanical Gardens, and by Rose in the Arthur Herbarium, Perdue University, USA (Pavlich Herrera, 1976). More recently there have been collections in the Selva regions by the Botanical Gardens of the University of Missouri-Columbia which has a permanent station in Oxapampa. The extent and composition of these collections is not known.

Most of the more recent national literature on macrofungal diversity in Peru comes not from Andean regions but from guides of macrofungi of the eastern Selva and lower Amazon forest regions such as Allpahuayo-Mishana, Iquitos (Espinoza M., 2003; Mata et al., 2006) and Madre de Dios (Alvarez Loayza et al., 2014; Cárdenas Medina et al., 2019; García Roca, 2015; Gazis, 2007; Sourell et al., 2018) and the Selva of Júnin (Salvador Montoya 2009). It is unclear to what extent some relied on physically collected and carefully processed samples, or photographic records and best guesses. Unfortunately, transparency of the identification process is lacking in most guides.

Knowledge of fungal diversity in highland Andes relies principally on the collections of Herrera and Pavlich and recently by collections in Cusco by Franquemont (Franquemont et al., 1990), Maria Encarnacion Holgado (Holgado Rojas et al., 2007) and others (Quispe et al., 2006; Chimey Henna and Holgado Rojas, 2010; Holgado Rojas et al., 2010; Trutmann, 2012; Quispe Pelaez, 2020). In the northern Andes information is available from a collection by the National University of Trujillo obtained by a recent thesis study of macrofungi of Andean Piura (Palacios Noe, 2015), as well as a guide of macrfungi of Cañaris in Lambayeque that came out of this study (Trutmann et al., 2019).

Added to these there is a guide to edible fungi in Peru that was published by Pavlich (Pavlich Herrera, 2001) and a very limited overview of edible fungi of Peru published by the Food and Agriculture Organization of the United Nations (FAO) as part of a global assessment of wild edible fungi (Boa, 2004), as well as a list of psychoactive fungi published by Gastón Guzmán and colleagues (Guzman, 2000).

Purpose and limitations

Development of primary knowledge by collecting and registering primary information of individual Andean macrofungi from the north to the south of Peru became the objective of this initiative of Global Mountain Action (GMA). Its purpose was to establish baseline information for future research on the importance, education, awareness of macrofungi in Peru and provide information to support the Peruvian government's ability to develop enabling environmental policies in mountains that include fungi.

This compendium documents the outcome of this work. It is a first effort to obtain an Andean wide idea of macrofungal diversity to which future (and past) collections could be added. Due to the size and the taxonomic scope of the collection and limited resources, it has not been possible or not intended to provide detailed taxonomic identifications of the 750 fungi collected. Instead, its purpose is to provide basic visual and written macroscopic and microscopic descriptions to support future taxonomic work, either through access to herbarium specimens, or the GPS locations to conduct specific collections as often only one specimen of a fungus was found. Its other purpose is to provide material for environmental awareness campaigns, education and related mycological research.

Finally, despite the mentioned limitations of the compendium the work is significant and important if only for the unique combination of visual illustrations, location data and generally Genus and morphospecies level identification of each sample that provides a user-friendly entrance into understanding the substantial diversity of Peruvian Andean mushrooms. It can be a guide not only for professionals in determining scientific studies, but also for amateur mycologists and eco-tourists since the compendium can be divided into regional, or even local guides containing visual and descriptive information of local fungi.

Taxonomic organization

The identification of members of genera used in this compendium follows, to the best of knowledge, the current accepted convention. However, there is much flux in the taxonomic status of macrofungi due important new molecular insights published all the time. Due to the limited access to the literature during most of the time of writing some of these current changes may have been missed. Numerous identifications are still not fully certain as they simply did not key out in the taxonomic literature available. Perhaps because they are still undescribed species, or because specific important features for identification to Genus, or species, were not noted during the broad-based diagnostic observations of bulk collection samples soon after processing.

To understand the contextual basis of the collection the macrofungi are presented in hierarchical form in the following manner. The Kingdom Fungi will be shown as the Phylum's Ascomycota and Basidiomycota. Within the Basidiomycotina they are placed in the Class <u>Agaricomycotina</u>, <u>Pucciniomycotina</u> and <u>Ustilaginomycotina</u>.

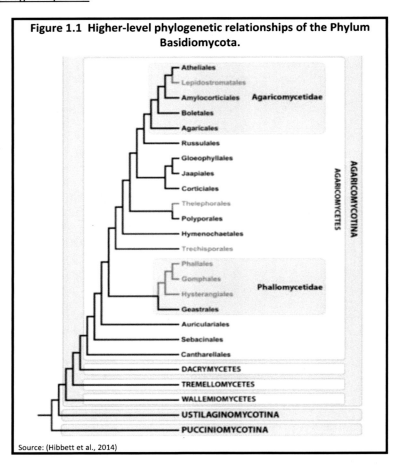

Figure 1.1 Higher-level phylogenetic relationships of the Phylum Basidiomycota.

Source: (Hibbett et al., 2014)

This study was principally concerned with non-microscopic representatives of the Basidiomycota, and to a lesser extent the Ascomycota. The Basidiomycota collected are in the Class Agaricomycotina, in particular the <u>Agaricomycetes</u>, <u>Dacrymycetes</u> and <u>Tremellomycetes</u> (figure 1.1). The most predominant Clade containing macrofungi is the Agaricomycetes (Table 1). The Clade Agromycetes contains 21 Orders (Table 1.1). The collection contains representatives all but the Atheliales, Gomphales, Jaapiales, Sebacinales, Trechisporales. Families and Genera in each group will be discussed separately in each group together with the collected samples. The Clade Dacymycetes contains 10 Orders. The Clade Tremellomycetes contains 3 Orders. Each is represented by at least one Order in Peru. Each Order will be treated separately in the compendium with an introduction, a list of Families. Each of the Families will also be introduced separately with information on each Genus. Members collected and analyzed of each Genus are presented in alphabetical order.

Table 1.1	The Class Agaricomycotina	
Clade Agaricomycetes	Clade Dacrymycetes	Clade Tremellomycetes
Sub Class Agaricomycetidae		
Agaricales (32 fam., 410+ gen.)	Calocera	Cystofilobasidiales
Agaricales-Clavariaceae (Clavariales?)	Cerinomyces	Filobasidiales
Amylocorticiales (1 fam., 14 gen.)	Cerinosterus	Tremellales
Atheliales (1 fam., 22 gen.)	Dacrymyces	
Boletales (16 fam., 95+ gen.)	Dacryonaema	
Jaapiales (1 fam., 1 gen.)	Dacryopinax	
Lepidostromatales (1 fam., 3 gen.)	Dacryoscyphus	
	Ditiola	
Sub Class Phallomycetidae	Femsjonia	
Geastrales (1 fam., 8 gen.)	Guepiniopsis	
Gomphales (3 fam., 18 gen.)		
Hysterangiales (5 fam., 18 gen.)		
Phallales (2 fam., 26 gen.)		
Incertae sedis (no sub-class)		
Auriculariales (6–7 fam., 30+ gen.)		
Cantharellales (7 fam., 39 gen.)		
Corticiales (3 fam., 30+ gen.)		
Gloeophyllales (1 fam., 7 gen.)		
Hymenochaetales (3 fam., 50+ gen.)		
Polyporales (9 fam., ~200 gen.)		
Russulales (12 fam., 80+ gen.)		
Sebacinales (1 fam., 8 gen.)		
Stereopsidales (1 fam., 2 gen.)		
Thelephorales (2 fam., 18 gen.)		
Trechisporales (1 fam., 15 gen.)		
Source: Wikipedia		

The three volume compendium is a first overview and guide to macrofungi of the Peruvian Andes based as much as possible on current general taxonomic status of taxa using the available literature. When inadequate descriptive information to identify genera and species is provided readers are given herbarium and GPS sources to pursue further studies.

METHODOLOGY

Hundreds of locations in the Peruvian Andes were surveyed for macrofungi. They ranged from the north to the south of the Peruvian Andes in the provinces of Puno, Arequipa, Cusco, Apurímac, Lima, Junín, Cerro de Pasco, Ancash, Cajamarca, La Libertad, Lambayeque, Piura and Tumbes.   Collection of macrofungi was emphasized in the Cordillera Central and Oeste.   Most of the samples were collected at altitudes from 1,000m to around 4,500m above sea level. Sampling at each site followed a random scan approach of an area of approximately 50 x 50 meters, or in 20 to 30-minute time periods. Sometimes in forests the time was extended to encourage the collection of samples of all species in a given area. Location information was recorded using a N-1 GPS recording unit connected to a

Nikon camera and included longitude and latitude coordinates as well as altitude. Each sample was photographed from various angles using Nikon D300, D610, D7500 or D100 cameras and stored, each separately, in a paper bag with an identification code. Regions were sampled between one and three times in consecutive years.

The GMA and the 'P' identification code

Each sample has been assigned a unique collection identification number: 1. ORGANIZATION 2. MAIN COLLECTOR 3. YEAR OF COLLECTION 4. COUNTRY 5. SOUTH, CENTRAL or NORTH REGIONS 6. ANDES or COAST 7. LOCATION 8. SAMPLE NUMBER. For example: GMA PT 1 PSA 3.1 = GMA (Global Mountain Action) PT (Peter Trutmann) 1 (first year of collection starting 2011) P (PERU) S (South) A (ANDES) 3.1 (Location 3).

Identified samples were also given a taxonomic name with annotated confidence levels and a numbered identification code for easy reference beginning with 'P' for Peru (e.g. P125), which serves to denote a morphospecies in the absence of certain bionomial taxonomic identification to the Genus or species level. This 'P' code is given on the top left of each illustration and is useful for identification and grouping when naming is uncertain. Below the 'P' number is given the number of collection samples attributed to that 'P' accession when greater than 1.

Sample processing

Fungi were measured, described, and processed as soon as possible after collection while the samples were fresh. These included basic fruiting body characteristics such as size, color, texture, substrate etc. (Largent 1974). They were prepared for spore prints by placing the hymenal side of the caps or structures on black and white paper and left covered overnight. Spore print color was recorded early the next day before further collection activities. The other parts of the samples were dried as soon as possible. The first year, in the absence of electric drying equipment, samples were air-dried in brown paper collection bags. This method was problematic especially in places like Puno with high humidity and low temperatures. In subsequent years samples were dried in the evenings in a portable electric dehydrator (Nesco Professional 700 watts at 35°C under continuous air flow). The samples were dried for 8-12 hours (mostly at night). In the morning, the dried samples were placed back in their original paper bags. At the earliest opportunity, the dried samples were frozen for 12-24 hours to eliminate associated microfauna such as mites, larvae, and eggs within the limitations of traveling. Despite the care and protection, especially in the north, numerous samples were lost due to insects and their own decomposition in the hot local climatic conditions. On each return to the city of Lima, all samples were frozen again for at least 48 hours to eliminate any surviving fauna and their eggs and dried for prolonged periods between 48 and 72 hours or until the material was completely dry.

Microscopic analysis was carried out immediately after drying and continued until all samples of the year were processed. In the first year the work was carried out in the mycology laboratory of the Faculty of Sciences of the Universidad Peruana Cayetano Heredia (UPCH) in the city of Lima, with the support and guidance of taxonomist Dr. Magdalena Pavlich, using a Zeiss stereomicroscope and an Olympus CH2 with B145 head compound stereomicroscope. In subsequent years, to speed up identification and evaluation, samples were observed using a simpler Olympus-GB compound microscope at the Global Mountain Action facilities in Lima. Samples were treated before viewing with potassium hydroxide (KOH) and Melzer's reagent and rarely with other reagents. During an evaluation in the first two years, observations focused where possible on spore, basidia, cystidia and pileipellis descriptions of frequently deteriorated material caused by suboptimal conditions during collection. But in subsequent years the samples were generally subjected to a standard set of observations that included characterizations of spores, basidia, cystidia, pileipellis as well as the hymenophore trama. Although useful to process bulk samples this strategy did not take into account highly specific characteristics required to distinguish members of various taxonomic groups. In all years, due to lack of inbuilt cameras on compound microscopes, all images were taken simply with a handheld Nikon Coolpix X 12 Megapixel camera placed on top of the compound microscope objective lens. This provided for less than perfect images, yet sufficiently high quality to document important microscopic characters of each sample.

Division of samples to a national institution

Each sample was divided, depositing half in paper bags for the climatized herbarium of the Universidad Peruana Cayetano Heredia (UPCH) through the Mycology Laboratory of the Faculty of Sciences. On arrival samples were again frozen for an extended period. In the case of the Universidad Nacional San Antonio Abad de Cusco (UNSAAC), the collections were made directly by them, therefore there was no need to provide samples of the samples collected since we had no control of their inventories respecting the procedures and autonomy of the UNSAAC. The samples kept in the herbarium at the UPCH are currently waiting to be processed further to incorporate repack and add morphological descriptions, and to recode them with UPCH's own collection numbers.

Identification

The collection was generally characterized taxonomically with various degrees of certainty only down to the Genus level in order to aid in the selection of taxonomic groupings of specimens for more detailed taxonomic work in the future when additional financial resources and taxonomic expertise are secured. The procedure described was selected as the best strategy to efficiently process the large number of samples collected, including those with unknown taxonomy. All samples were described morphologically and supported with geographical and ecologically information as described above, and so placed in initial taxonomic groupings as best as possible to the Genus level. Main taxonomic resources used were those of Richard Dennis (1970), Ralf Singer (1975), David Arora (1986), and Thomas Laessoe and Jens Petersen (2019). These were supported by available mushroom guides from Peru and surrounding countries and internet resources. Molecular characterization of the collection was not accessible at the time.

AN OVERALL SUMMARY OF FINDINGS

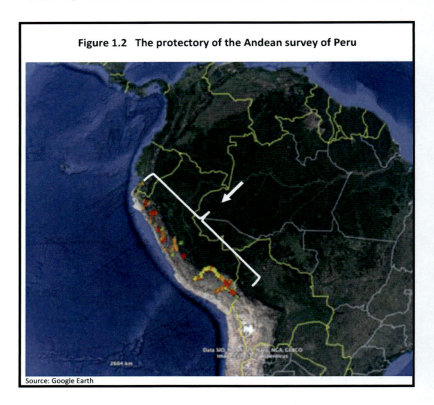

Figure 1.2   The protectory of the Andean survey of Peru

Source: Google Earth

Coverage of the Peruvian Andes

The full survey and collection documented in Parts 1 to 3 of 'Mushrooms of Andean Peru' covered most of the Central to Western Peruvian cordilleras Andes from north to south of Peru, a distance spanning more than 2500 km of topographically highly variable landscapes (figure 1.2). Many sites in the south (Puno and Cusco), center (Ancash) and north (Lambayeque) were sampled for three consecutive years, others at least once. Here is a preliminary analysis of the full collection given as prelude to a detailed analysis that will be published in the future.

Samples collected

Table 1.2 Number of samples collected		
Samples collected and evaluated	Samples collected not yet evaluated or without images	TOTAL
684	44	728

The collection provides more complete baseline information about Peru's unique Andean fungal diversity to support national environmental policy makers, research institutions, indigenous and civil society groups, as well as international organizations to strengthen national and international efforts to implement the General Environmental Law No. 28611 to document and help conserve the biological diversity of mountain ecosystems as established in articles 93 and 100.

In total, around 730 macrofungal samples were collected (table 1.2). Of these samples, 94% (684) have been evaluated macroscopically and microscopically and have in situ field images that were used to document the diversity in publications. A small proportion of the collection representing 6% (44 samples) has not been fully evaluated due to lack of field images. The collection provides a national baseline from north to south of the Andean region of Peru to which other collections can be added and compared. Added to these, the full compendium includes 133 additional photographic records and descriptions of samples lost for various reasons, or that were not collected physically due to other limitations. These samples have not been used for the analyses that follow.

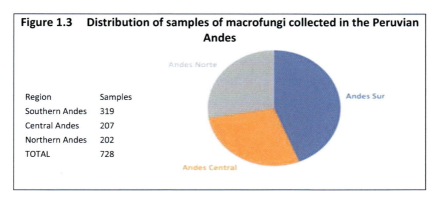

Figure 1.3   Distribution of samples of macrofungi collected in the Peruvian Andes

Region	Samples
Southern Andes	319
Central Andes	207
Northern Andes	202
TOTAL	728

The collection's representativeness is slightly skewed to the south of the Andes (figure 1.3). About 40% of the samples in the herbarium are from the south, while about 30% are from the central Andes and about 30% from the northern Andes.

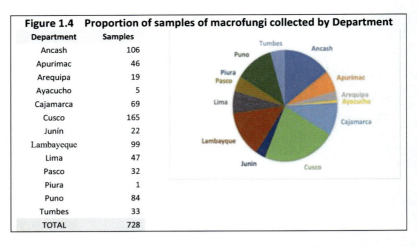

Figure 1.4 Proportion of samples of macrofungi collected by Department

Department	Samples
Ancash	106
Apurimac	46
Arequipa	19
Ayacucho	5
Cajamarca	69
Cusco	165
Junín	22
Lambayeque	99
Lima	47
Pasco	32
Piura	1
Puno	84
Tumbes	33
TOTAL	728

Most of the Andean regions are represented in the collection; however, representation by Department is very uneven. Some regions are much better represented in the collection than others (figure 1.4).   Samples from Cusco and Puno most prominently represent the southern Andes, samples from Ancash dominate the central Andes, and samples from Cajamarca and Lambayeque are most numerous in the northern Andes. The Andean regions of Arequipa, Ayacucho and Piura are poorly represented in this collection. Although all regions require follow-up collections, the mentioned regions require priority in future collections of the diversity of Andean macrofungi.

Taxonomic diversity of the Andean collection

The collection contains representatives of both Basidio- and Ascomycota, although the Basidiomycota were emphasized during the search. In 684 samples, 7 Classes, 17 Orders, 51 families and 162 genera and 513 morphospecies were identified (table 1.3). This number is likely to change slightly in the future when the collection is evaluated more thoroughly.

Table 1.3 Taxonomic categories in the collection (n=684)

Phylum	Class	Orders	Families	Genera	Morpho-species	Total samples
Basidiomycota	3	11	42	150	513	664
Ascomycota	5	6	9	12	16	20
Total	**7**	**17**	**51**	**162**	**529**	**684**

There are few published studies in South America that enable these data to be compared with respect to diversity. One such study was published by López

Quintero et al (2012). They carefully studied the macrofungal diversity in forests of two Colombian Amazon regions 300 km apart. From a total of 888 samples collected, they identified 403 macrofungal morphospecies belonging to 129 genera and 48 families of Basidiomycota and Ascomycota. Given that the study was conducted in the Amazon region which is considered by many to be the most diverse in the world, as a measure of diversity our collection results compare well and demonstrate the richness of diversity in the Peruvian Andes. In fact, in this study using the frequency of collection information (i.e. the number of samples needed to collect a new Family or Genus), it is clear that only a small fraction of the existing diversity of species in the Andes has been collected to date.

Fungal diversity in different ecosystems

A first look at the collection's taxonomic data points to substantial differences in the genetic diversity of macrofungi in different ecosystems and land use systems in the Andes as measured by the presence of families and genera (Table 1.4). Even with only a relatively small sample base that may only represent a few percent of the actual diversity, the data shows the presence of rich biodiversity, especially in the northern cloud forests. These cloud forests harbor highly threatened and unique ecosystems (Llatas-Quiroz and López-Mesones 2005). A lower level of macrofungal diversity was found in the puna of the altiplano and natural pasture systems. The diversity of macrofungi in introduced Pinus and Eucalyptus plantations was surprisingly high.

Table 1.4 Initial insights into the taxonomic diversity of macrofungi between Andean land use systems			
		Taxonomic category	
Management system	Altitude (meters)	Families /100 samples	Genera /100 samples
Puna and natural highland pastures	2993-4650	11	28
Polylepis cloud forests	2900-4057	19	43
Northern relic Amazonian cloud forest	2293-2803	35	57
Pinus and Eucalyptus plantations	2706-4063	26	48

Representativeness of collection of total macrofungal diversity in ecosystems sampled

It is sobering that to date the data indicates we only collected a very small percentage of the total macrofungal diversity of various regions, depending on location. For example, Andes-wide depending on ecosystem, site and year, we collected a new Genus at the numerous collection sites every 1.1 to 2.6 samples and a new Family every 2 to 4.5 samples. Thus, there is plenty of room for aspiring mycologists to fill in the knowledge gap. This task should become ever easier with new molecular tools and field equipment.

From a global perspective, according to Hawksworth, the number of fungi on earth range from 500,000 to 9.9 million species, of which only 120,000 are named. A working figure of 1.5 million species is generally accepted (Hawksworth, 2001). Curiously, this estimate still appears to be the most authoritative 15 years later. Most new mushrooms are being discovered in the tropics, especially those forming ectomycorrhizas with native trees. In various tropical areas 73% of mushroom species were shown to be undescribed. Moreover, collections made over periods of a few years or less underestimate the actual species present. Many morphologically defined mushrooms prove to be assemblages of many biological species; the existence of cryptic species means that the number of known species may be an underestimate by a factor of at least five. The number of mushrooms on earth is estimated at 140,000, suggesting that only 10% are yet known. This figure was calculated by extrapolation of the proportion of mushrooms in the known fungi (18.75%) to the overall 1.5 million species estimate, with reductions to allow for the extent of novelty actually being found and for a conservative allowance for numbers of cryptic species. It suggests that in the Andes we have thus far severely under-collected and underestimated the actual diversity of macrofungi.

Challenges of collecting in the Andes

Although more than 728 samples were collected and stored, a substantial number of them were lost due to their fragility and biological composition, and because they could not be stabilized prior to decomposition by microorganisms and consumption by associated microfauna. Where possible images and descriptions of these were used in the compendium. Most of the fruiting bodies of macrofungi are fragile and deteriorate rapidly, which is why they require immediate conservation processes and technology. Collection sites are far from main roads as are facilities that enable drying and freezing samples. Another challenge in the collection was the highly variable weather and rainfall during the collection years that affected the quantity and quality of the samples and the organization of the collection trips. For example, 2015 was a dry year when mostly over-mature samples were found. These were difficult to conserve and also less than desirable as herbarium specimens. Conversely, 2017 was a year of excessive rains and floods that prevented travel and collection. As a result, some regions such as Piura and Ayacucho are under-represented in the collection. One year, theft of a substantial part of our equipment and damage to the vehicle affected the ability to collect.

Using the compendium of the collection

In the parts that follow, individual samples of macrofungi from the collection are described and identified. Readers are provided with an easy overview section where mushrooms likeness can be visually scanned and then checked more carefully using the 'P' number by going to the respective page. The collection is presented in alphabetical order respecting certain taxonomic categories. That said, some categories were adjusted to fit knowledge of the materials. It is hoped the format provides a useful and easy way to manage the material. At the end there is a reference section and a collection number index.

Finally, we hope that the compendium will be informative and for the reader a magic journey of discovery of each fascinating fungal individual discovered living in the Peruvian Andes.

Viva los hongos!

BASIDIOMYCOTA

AGARICOMYCOTINA
Agaricomycetes
Sub Class Agaricomycetidae
Order Agaricales

The Agaricales or Euagarics clade (Basidiomycota, Agaricomycetidae) is the largest clade of mushroom forming fungi and includes more than half of all known species of the homobasidiomycetes (Matheny et al., 2006). There is still disagreement on its taxonomic composition. The Order proposed by Matheny et al (2006) will generally be followed in this compendium. The Order has 33 extant families, 413 genera, and over 13,000 described species, along with six extinct genera known only from the fossil record (figure 1.5).

Figure 1.5   Cladeogram of the Agaricales, showing the division into six major clades

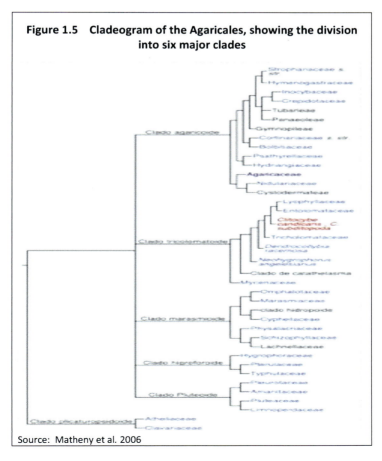

Source: Matheny et al. 2006

Family Agaricaceae

Tribe Agariceae Singer

© Peter Trutmann

Agaricus

Section Agaricus-like:
Not staining when damaged, annulus thin and slight or even absent KOH negative (Arora 1986:313)

P1   ***Agaricus* cf. *bisporus***   Spores: x̄ (5.8x4.2)μm (Qm = 1.4)
              Southern Andes: Puno

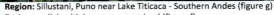

Region: Sillustani, Puno near Lake Titicaca - Southern Andes (figure g)
Ecology: soil, in altiplano puna grassland (figure f)

Macroscopic
Odor: pleasant (sweet)
Color reaction: No change to yellow or red
Pileus: (5-8)cm, convex to plane, white, smooth, rugose, fleshy, dry, dull (fig. a)
Stipe: 3-4(1.5)cm, 1.5cm wide, white, fleshy, with white scales and reduced, delicate membranous annulus (figure b)
Gills: pink to brown, free, crowded (figure b)
Spore print: fuscous brown

Microscopic
Spores: (5.0-7.0)x(4.0-5.0)μm, x̄ (5.8x4.2) μm (Qm = 1.4) (n = 7), oval, smooth, brown (KOH), without germ-pore (figure c)
Trama: parallel (figure d: black arrow)
Basidia: clavate, with two sterigmata (figure e: white arrow)
Pileipellis: a cutis, with hyphae parallel to the surface
Sample: GMA PT 1 PSA 9.10
3876 m 15°43'19.865" S 70°9'22.536" W
Observations: *Agaricus* (Singer 1975:458): conforms to description of *A. bisporus* including having two spored basidia and spores (5.5-8.5)x(4-6.5)μm (Arora, 1986). In South America reported in Argentina by Wright y Albertó (Wright J.E. Albertó E., 2002: 208) and Cusco, Peru (Quispe pers com.).

Agaricus cf. pampeanus

Spores: x̄ (8.2x5.4)μm (Qm=1.5)

Central and Southern Andes: Cerro de Pasco, Junín, Puno

Region: Cerro de Pasco, Junín and Puno - Central and Southern Andes (figure h)
Ecology: soil in highland puna grassland (figure h)
Macroscopic
Odor: very mild
Color reaction: no
Pileus: 3-6 cm, convex, smooth, white-colored often with darker streaks (figs. a,b)
Stipe: 1.5-3 (x2) cm, white, with delicate ring (figure c)
Gills: pink-brown, free, crowded (figure d)
Spore print: fuscous brown
Microscopic
Spores: (7.5-10)x(5.1-7)μm, x̄ (8.3x5.4) μm (Qm = 1.5) (n = 10), ovoid-ellipsoid, brown(KOH), smooth, with small germ-pore (figure d)
Cystidia: not found (figure d)
Trama: parallel (figure g: white arrow)
Basidia: (24)x(6-7.5) μm, 4 spores (figure e)
Pileipellis: a cutis: of mycelium parallel to surface (figure f: black arrow)
Samples:
Yauli, Junín   GMA PT 7 PCA 2.1
4613 m   11°16'53.49" S 76°26'47.292" W
                GMA PT 7 PCA 2.4 ,
4526 m   11°6'13.182" S 76°22'49.914" W
Pasco, Cerro de Pasco: GMA PT 7 PCA 2.7B
4650 m   11°2'26.969" S 76°23'36.54" W
Lampa, Puno: GMA PT 1 PSA 1.5,
3852 m   15°18'47.153" S 70°12'41.279" W
Hacienda Icshuya, Puno: GMA PT 1 PSA 0.1
4396 m   15°43'51.503" S 70°50'57.215" W
Observations: *Agaricus:* (Singer 1975:458) . Like *Agaricus pampeanus Speg* with spores (7.5-11x5.2-7.2)μm (Wright J.E. Albertó E., 2002:214).

5 μm
1 unit = 1.5 μm
1 unit = 3.75 μm
25 μm

Agaricus sp.
Central Andes: Lima

Spores: x̄ (8.4x5.8)μm (Qm=1.4)

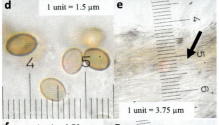

Region: Canta, Lima (figure i)
Ecology: soil in agro-pasture (figure h)
Macroscopic
Odor: pleasant
Color reaction: none
Pileus: 3.5-8cm, plane, white with slight browning and streaking, smooth, dry (figures a and b)
Stipe: 1 -1.5cm, white, with white scales below intermediate type annulus (figure a,b)
Gills: pink to chestnut, free, crowded (figure c)
Spore print: fuscous brown
Microscopic
Spores: (7.5-9)x(5.5-6)μm, x̄ (8.3x6.0)μm (Qm=1.4) (n = 10), ovoid - ellipsoid, brown(KOH), smooth with narrow germ-pore (figure d)
Trama: regular (parallel) (figure g: red arrow)
Cystidia: not found
Basidia: short (22-26) x (7.5) μm with 4 spores (figure g: white arrow)
Pileipellis: a cutis of mycelium parallel to surface(figure e: black arrow)
Sample: GMA PT 7 PCA 1.3
3343 m   11°27'16.667" S 76°35'50.364" W
          GMA PT 7 PCA 1.20
3607 m   11°26'29.784" S 76°35'7.59" W
Observations: *Agaricus* (Singer 1975:458). Very similar to *A.pampeanus*, but with more slender stipe and found at lower altitude than P2.

P4 ***Agaricus sp.***

(2) Central Andes: Ancash

Spores: x̄ (8.3x6.0)μm (Qm=1.4)

Region: Yungay, Ancash -Central Andes (figure h)
Ecology: agro-pastures (figure g)
Macroscopic
Odor: slightly tar like (fresh) very agreeable (dry)
Color reaction: no yellowing
Pileus: 8cm, plane to broadly convex with marked sides, white, smooth, shiny, dry, often with slight depression (figures a and b)
Stipe: 5-6 (1-1.5) cm, white, smooth, with a fragile membrane type annulus (figure c)
Gills: pink to dark brown, free, crowded (figure c)
Spore print: fuscous brown (figure c: orange arrow)
Microscopic
Spores: (7.5-9.4)x(4.9-7.5) μm (8.3x6.0)μm (Qm=1.4) (n=10), oval-elliptical, brown (KOH), smooth, without germ-pore (figure d)
Trama: parallel (figure f: white arrow)
Cystidia: not found
Basidia: (18-24) x (4-6) μm, generally with two sterigmata (figure e: black arrow)
Pileipellis: not evaluated
Sample: GMA PT 3 PNA 8.9A,
2511 m   9°13'8.675" S 77°41'11.406" W
                GMA PT 3 PNA 8.9B,
2506 m   9°13'8.52" S 77°41'11.268" W
Observations: *Agaricus* (Singer 1975: 458). With smooth cap and larger spores than *A.campestris* (5.5-8.5x4-6.5 μm), and cap shape and appearance different to *A.pampeanus*. Traditional local name: 'Tuklla'

1 unit = 3.75 μm

P5 *Agaricus sp.*
Spores: x̄ (5.6x4.4)µm (Qm=1.3)
Central Andes: Ancash

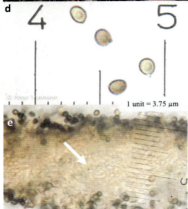

Region: Recuay, Ancash - Central Andes (figure g)
Ecology: organic matter, below native trees close to the river (figure f)
Macroscopic
Odor: delicious organic smell (fresh)
Color reaction: none
Pileus: 3-7 cm, white, plane, matt, powdery, with slightly rough edges (figure a and b)
Stipe: 4-6 (0.5-1) cm, white with powdery white scales, with very faint membranous annulus figure b)
Gills: brown, free and crowded (figure c)
Spore print: fuscous brown (figure c: orange arrow)
Microscopic
Spores: (5.2-6)x(4-4.5) µm, x̄ (5.6x4.4)µm (Qm=1.3) (n=5), elliptical-oval, brown with a central vacuole (KOH), smooth, no germ-pore (fig. d)
Trama: interwoven (figure e: white arrow)
Cystidia: not observed
Basidia: not evaluated
Pileipellis: not evaluated
Sample: GMA PT 3 PNA 9.19,
1671 m 10°9'20.118" S 77°31'19.47" W
Observations: *Agaricus* (Singer 1975: 458): Enjoyed by larvae

P6 *Agaricus sp.*
Central Andes: Ancash

Spores: x̄ (5.0x4.1)µm (Qm=1.2)

Region: Recuay, Ancash - Central Andes (figure g)
Ecology: organic material below native tree near the river (figure g)
Macroscopic
Odor: like Portobello (fresh, tarlike (dry)
Color reaction: no
Pileus: 6-7cm, broadly convex, in-turned margin, flat top, dry, shiny, white-cream with fine white scales browning with age (figs. a,b).
Stipe: 6-7.5cm, white-cream with white scales, remnants of an membranous intermediate (perhaps even sheath type) ring (fig. b)
Gills: dark brown, free, crowded (figure c)
Spore print: fuscous brown (figure c: orange arrow)
Microscopic
Spores: (4.5-5.6)x(4.0)µm, x̄ (5.0x4.1)µm (Qm=1.2) (n=5), small, spherical to oval, dark brown (KOH), smooth, no germ-pore (figure d).
Trama: parallel (figure f: white arrow)
Cystidia: not found
Basidia: 2-4 sterigmata, 15-20µm (figure e: black arrow)
Pileipellis: not evaluated
Sample: GMA PT 3 PNA 9.21
1664 m   10°9'20.496" S 77°31'19.524" W
Observations: *Agaricus:* (see Singer 1975:458)

1 unit = 3.75 µm

P7 *Agaricus campestris*
(3)   Southern Andes: Cusco

Spores: x̄ (6.9x4.9)µm, (Qm=1.4)

1 unit = 1 µm

1 unit = 2.5 µm

Region: Canchis and Cusco, Cusco - Southern Andes (figure f)
Ecology: organic matter-soil associated with agro-pastures with young *Polylepis* trees at historic site (figure a-c)
Macroscopic
Odor: agreeable. Taste: agreeable
Color reaction: none
Pileus: 4-6 cm, white, squamous with cream to brown scales, dry, (fig. a,b)
Stipe: 2-3 (1-2) cm, white to cream, intermediate membranous type annulus
Gills: pink to dark brown, free, crowded (figure c)
Spore print: fuscous brown
Microscopic
Spores: (6.0-8.0)x(4.0-6.0)µm, x̄ (6.9x4.9)µm, (Qm=1.4) (n=10), elliptical-oval, brown (KOH), smooth, without germ-pore (figure d)
Cystidia: not found
Basidia: (22x8)µm, 4 spored
Pileipellis: a cutis (figure e: black arrow)
Sample:
Canchis, Cusco: GMA PT 1 PSA 7.2
3500 m   14°10'24.431" S 71°22'8.549" W
Cusco, Cusco: GMA PT 2 PSA 5.1,
3024 m   13°34'9.33" S 71°51'57.66" W (5.1)
Cusco, Cusco: GMA PT 2 PSA 7.4
               13°31'9.229" S 71°55'51.989" W)
Observations: *Agaricus* (Singer 1975:458): like *Agaricus campestris* Linnaeus (6.5) 7-8 x 4.0-5.5µm (Wright J.E. Albertó E., 2002:210.) (Wright J.E. Albertó E., 2002: 210), Arora (1986:318). Reported from central Peru (Pavlich Herrera, 1976) and Cusco (Quispe et al., 2006). Samples consumed by insects. Traditional edible Quechua 'Kallampa'

P8 *Agaricus sp.* Spores: x̄ (5.8x 3.9)μm (Qm=1.5)
Southern Andes: Apurimac

Region: Abancay, Apurimac (figure e)
Ecology: organic matter in *Podocarpus glomeratus* forest (figure d)
Macroscopic
Odor: N.E.
Color reaction: not staining or slightly reddish?
Pileus: 6-7cm, plane-umbonate, white, squamous with brownish-red scales, darker at umbo, dry (figure a)
Stipe: 9cm, white, slender, smooth, delicate, intermediate type annulus (figure b)
Gills: pink to brown, free, crowded (figure b)
Spore print: dark brown
Microscopic
Spores: (5.2-6.7)x(3.7-4.2)μm, x̄ (5.8x 3.9)μm, (Qm=1.5) (n=6), elliptical, brown (KOH), smooth, without germ-pore (figure c)
Sample: GMA PT 3 PSA 2.38
3273 m   13°35'32.076" S 72°52'53.886" W
Observations: *Agaricus* (Singer 1975:458)

1 unit = 3.75 μm

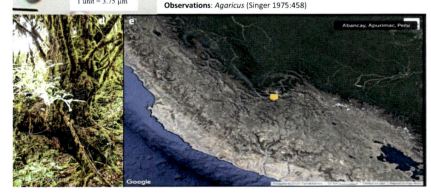

Agaricaceae - cf. Agaricus Spores x̄ (5.5x3.7)µm (Q=1.5)
Northern Andes: Lambayeque

© Peter Trutmann

Region: Salas, Lambayeque, Lambayeque (figure g)
Ecology: Leaf litter under native tree (figure f)
Macroscopic
Odor: pleasant
Color reaction: none
Pileus: 3.5-4 cm, plane, delicate, white with grey-brown, squamulose
with punctuate scales and a darker center, dry (figure a)
Stipe: 10-11cm, white, smooth with remnants of a delicate
membranous intermediate annulus (figures a,c)
Gills: brown, free, crowded (figure c)
Spore print: fuscous brown
Microscopic
Spores: (5.6-7.1)x(3.7) µm, x̄ (5.5x3.7) µm (Qm=1.5) (n=5) elliptical
smooth, dark brown (KOH), without germ-pore (figure d)
Trama: not evaluated
Cystidia: not found
Basidia: with 2 or more spores (figure e: black arrow)
Pileipellis: not evaluated
Sample: GMA PT 3 PNA 1.4
1162 m 6°13'31.037" S 79°31'39.786" W
Observations: *Agaricus* (Singer 1975: 458): not in Dennis (Dennis, 1970:
58), Neves et al (Neves et al., 2013), Franco-Moleno et al (Franco-
Molano A.E., 2005), Mata et al (Mata et al., 2006) or Alvarez et al
(Alvarez Loayza et al., 2014)

1 unit = 3.75 µm

Section Hortensis-like:

Not staining or rufescent, annulus skirt-like or intermediate, stature normally rather squat, and KOH negative (Arora 1986:313)- pleasant odor

P10
(2)

Agaricus sp.
Southern Andes: Cusco

Spores: x̄ (7.2x4.9)μm (Qm1.5)

1 unit = 3.75 μm

Region: Cusco, Cusco - Southern Andes (figure e)
Ecology: soil, agro-pasture (figure d)
Macroscopic
Odor: agreeable, taste: agreeable
Color reaction: none
Pileus: 5-6 cm, broadly convex, white, squamulose with small punctate brown scales, dry, silky (figure a)
Stipe: 3-4 (1) cm, white, smooth, intermediate to skirt type annulus (figure b)
Gills: Pink to dark brown, free, crowded (figure b)
Spore print: fuscous brown
Microscopic
Spores: (6.8-7.9)x(4-2-5.6) μm, x̄ (7.2x4.9)μm (Qm1.5)(n=5), ellipsoid-oval, brown (KOH, smooth, without germ-pore (figure c)
Cystidia: not evaluated
Basidia: not evaluated
Pileipellis: not evaluated
Sample: GMA PT 2 PSA 4.5,
3059 m   13°31'44.969" S 71°44'26.01" W
GMA PT 2 PSA 5.17
3589m   13°34'20.616" S 71°51'20.454" W
Observations: *Agaricus* (Singer 1975:458): similar to *A.campestris,* but with punctate scales on pileus with annulus tending toward a skirt.

P11 *Agaricus cf. nivescens*   Spores: x̄ (5.8x4.6)µm (Qm=1.3)
Southern Andes: Puno

Region: El Callao, Puno - Southern Andes (figure f)
Ecology: soil, in altiplano puna grassland (figure e)
Macroscopic
Odor and taste: agreeable
Color reaction: none
Pileus: 3-5cm, convex with in-turned edges or flat topped,
squamous - white with light brown scales especially when
young, dry, dull (figure a)
Stipe: 2(1)cm, membranous intermediary annulus (figure a)
Gills: pink to brown, free, crowded (figure a)
Spore print: fuscous brown
Microscopic
Spores: (5-7.5) x (3.5-5) µm, x̄ (5.8x4.6) µm (Qm=1.3) (n=3),
sub-globose, brown (KOH), smooth, no germ-pore (figure b)
Trama: parallel (figure c: white arrow)
Cystidia: none found.
Basidia: short with 2-4 spores (figure d: red arrow)
Pileipellis: a cutis: mycelium
Sample: GMA PT 1 PSA 8.4
3,873 m   16°5'51.629" S 69°37'8.141" W
Observations: *Agaricus* (Singer 1975;458): probably *Agaricus
nivescens (Müller)Möller*: Como (Wright J.E. Albertó E., 2002)
p.214. *A.nivescens* spores measure (5-7)x(4-5)µm.

P12
(2)
Agaricus sp.
Central Andes: Ancash

Spores: x̄ (6.5x4.2)µm (Qm=1.5)

Region: Yungay, Ancash - Central Andes (figure h)
Ecology: organic material under _Polylepis_ forest (figure h)
Macroscopic
Odor and taste: mild (O) agreeable flavor (T)
Color reaction: No color change
Pileus: 3-10 cm, convex when young to plane when mature, squamous with small to medium dark brown scales above white base (figures a,b)
Stipe: 4-6 (1) cm, white, with intermediary type annulus with short skirt - persistent (figure c)
Gills: pink to chestnut brown, free, crowded (figure d)
Spore print: fuscous brown (figure d: orange arrow)
Microscopic
Spores: (5.6-7.5)x(4-5)µm, x̄ (6.5x4.2)µm (Qm=1.5) (n=5), ovate to elliptical, brown when mature, smooth, without germ-pore (figure e)
Trama: parallel (figure f: black arrow)
Cystidia: not observed
Basidia: (c.19 x 6) µm, 4 spored (figure g: white arrow)
Pileipellis: cutis of unspecialized hyphae
Samples: Llanganuco, Yungay, Ancash
GMA PT 5 PNA 1.17,
3845m   GPS:9°4'41.472" S 77°39'8.31" W
GMA PT 5 PNA 1.20 ,
3919m   9°4'47.681" S 77°38'52.901" W
Observations: _Agaricus_ (Singer 1975:458): similar to _A.placomyces,_ but without yellow discoloration and without phenolic odor.

1 unit = 3.75 µm

Agaricus sp.

Southern Andes: Apurimac

Spores: x̄ (7.0x4.2)µm (Qm=1.7)

Región: Abancay, Apurimac (figure g)
Ecology: organic matter by track in a *Podocarpus glomeratus* forest (figure f)

Macroscopic

Odor: fragrant aromatic

Color reaction: none or faint red reaction

Pileus: 3.5-6cm, plane , sometimes split, from cream, light brown to grey when old, showing rugosity in the form of brown scales (figures a,b)

Stipe: 5-10 (1-1.5)cm, brown, with and short, white, short skirt-like annulus (figure c)

Gills: pink to dark brown, free, crowded (figure c)

Spore print: fuscous brown

Microscopic

Spores: (5.6-7.5)x(3.9-4.5) µm, x̄ (6.8x4.0)µm (Qm=1.7) (n=11), elliptical, brown(KOH), smooth without a germ-pore (figure d)

Trama: interwoven to parallel (figure e: white arrow)

Basidia: c.(12-16)x(4)µm, 2-4 spored (figure e)

Pileipellis: not evaluated

Samples: GMA PT 3 PSA 3.19
3273 m   13°35'32.076" S 72°52'53.886" W
            GMA PT 3 PSA 2.2
2948m   13°36'10.47" S 72°52'34.727" W

Observations: *Agaricus* (Singer 1975:458)

1 unit = 3.75 µm

P14
(2)

Agaricus sp.
Central Andes: Ancash

Spores x̄ (5.2x3.5)μm (Qm=1.5)

Region: Recuay, Ancash (figure j)
Ecology: organic debris, under native trees close to river (figure i)
Macroscopic
Odor: like Portobello (fresh) spiced (dry)
Color reaction: no or possibly slightly red
Pileus: 3-9 cm, convex (young) to plane (mature), white, slightly scaled, scales pink-brown to brown, center more densely colored (figure a, b, c)
Stipe: 6-9 cm, white, slightly scaled, membranous annulus (figures b,d)
Gills: pink to dark brown, free, crowded (figure d)
Spore print: dark brown (figure d: orange arrow)
Microscopic
Spores: small (4.5-5.6)x(3.5-4.5)μm, x̄ (5.2x3.5)μm (Qm=1.5)(n=10), subglobose-elliptical, brown (KOH), smooth, without germ-pore (figs e,f)
Trama: of large interwoven hyphae (figure g,h: black arrow)
Cystidia: not found
Basidia: pigmented (KOH) (figures g: white arrow), 4 sterigmata (figure f: red arrow)
Pileipellis: not evaluated
Sample: GMA PT 3 PNA 9.24,
1672m   10°9'19.949" S 77°31'18.414" W
          GMA PT 3 PNA 9.23B,
1660m   10°9'20.364" S 77°31'19.488" W
Observations: *Agaricus* (Singer 1975:458): separated from GMA 3 PNA 9.23B from the same location only by smell, spore print and its slightly more globose spore shape.

P15 *Agaricus sp.*
Central Andes: Lima

Spores x̄ (5.5x4.3)µm (Qm=1.3)

© Peter Trutmann

1 unit = 1.5 µm

1 unit = 6.5 µm

1 unit = 15 µm

© Peter Trutmann

Region: Zarate, Huarochiri, Lima - Central Andes (figure h)
Ecology: soil with moss below native shrubs (figure g)
Macroscopic
Odor and taste: agreeable, mild (O) agreeable (T)
Color reaction: none
Pileus: (6x6)cm, broadly conical with flat top to plane, white under squamulose layer of brown scales splitting to form radial lines, with darker center (figures a,b)
Stipe: 9cm, beige, smooth, central, with white, persistent, intermediate type annulus (figures b,c)
Gills: pink to chestnut brown, free, crowded (figure c)
Spore print: fuscous olive brown
Microscopic
Spores: (6-7.2)x(3.8-4.8)µm, x̄ (5.5x4.3)µm (Qm=1.3) (n=5), ovate, smooth, brown(KOH), inamyloid (Melzer's), no germ-pore (fig. d)
Trama: irregular-interwoven (figure f: red arrow)
Cystidia: not found
Basidia: (c. 14 x 4) µm, 2-4 spored, inamyloid
Pileipellis: a cutis: of parallel, repent hyphae - radial section (figure e: black arrow)
Sample: GMA PT 6 PCA 1.7
2936 m   11°55'52.47" S 76°28'59.43" W
Observations: *Agaricus* (Singer 1975:458)

THE MACROFUNGI OF ANDEAN PERU Part 1

Agaricaceae cf. *Agaricus*   Spores: x̄ (6.4x3.7)μm (Qm=1.7)
Southern Andes: Cusco

© Peter Trutmann   © Peter Trutmann

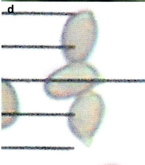

1 unit = 3.75 μm

Region: Ocra, Anta, Cusco (figure e)
Ecology: organic matter, pastures land with degraded native trees and Eucalypts (figure e)
Macroscopic
Odor: bleach or sperm-like
Color reaction: none (slightly brownish)
Pileus: 5 cm, lepiotoid, umbonate-plane, fragile, slightly floccose from pink scales above white base, with slightly shaggy margin (figures a,b)
Stipe: 5-6(0.5)cm, white to cream, fragile, with short skirt type membranous annulus, and peculiar foot (figure a)
Gills: pink to brown, free, crowded (figure c)
Spore print: did not yield print
Microscopic
Spores: (5.6-7.5)x(3.4-3.8)μm, x̄ (6.4x3.7)μm (Qm=1.7) (n=5), elliptical, brown(KOH), smooth without germ-pore (figure d)
Sample: GMA PT 3 PSA 5.13
3750 m   13°36'12.324" S 72°12'20.051" W
observations: *Agaricus* or *Micropsalliota* (Singer 1975:458): if it has pseudoamyloid spores with cheilocystidia then *Micropsalliota*, otherwise it is an *Agaricus sp.*

Provincia de Anta, Cuzco, Peru

Section Sanguinolenti-like:
Red staining, generally erect, with skirt like annulus, KOH negative with mild or non-phenolic odor

P17 **Agaricus sp.**   Spores: x̄ (7,3x4.0)μm (Qm=1.8)
Northern Andes: Cajamarca

1 unit = 3.75 μm

Region: Pucará, Jaen, Cajamarca - Northern Andes (figure f)
Ecology: soil on side of dirt road (figure e)
Macroscopic
Odor: earthy and pleasant
Color reaction: reddish
Pileus: 11cm, plane with upturned edges, white, fleshy, smooth, day (figures a,b)
Stipe: 5(1.5)cm, white, smooth, central, with membranous skirt annulus (figures a.c)
Gills: pink to brown, free, crowded (figure c)
Spore print: fuscous brown
Microscopic
Spores: (6.7-7.5)x(3.7-4.5)μm, x̄ (7,3x4.0) μm (Qm=1.8) (n=5) elliptical. smooth, brown
(KOH), without germ-pore (figure d)
Trama: not evaluated
Cystidia: not evaluated
Basidia: not evaluated
Pileipellis: not evaluated
Samples: GMA PT 3 PNA 2.1
1317m   6°1'28.265" S 79°12'13.967" W
Observations: *Agaricus* (Singer 1975:458): red coloration, pleasant odor and skirt annulus
suggests placement in the Section Sanguinolenti .

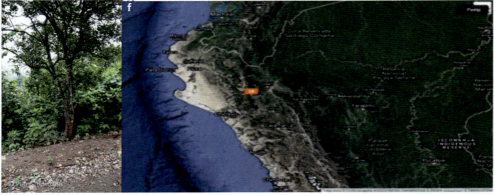

P18
(3)

cf. Agaricus sp.
Central Andes: Ancash

Spores: x̄ (6.4x4.0)µm (Qm=1.6)

© Peter Trutmann

1 unit = 3.75µm

Region: Llanganuco, Yungay, Ancash (figure h)
Ecology: organic litter, under *Polylepis* (figure h)
Macroscopic
Odor: agreeable like champignons
Color reaction: none (slightly browning)
Pileus: 9 cm, convex (young) to plane (mature), white, but when young
covered with smooth dark brown veil (figure b) that breaks as the cap
expands with brown scales (figure c) becoming more stringlike with a dark
center at maturity (figure a)
Stipe: 10 cm, white, lightly scaled and bulbous, with membranous,
intermediate type annulus (figure c)
Gills: pink to brown, free, crowded (figure c)
Spore print: no print
Microscopic
Spores: (5.6-7.1)x(3,8-4.5)µm, x̄ (6.4x4.0)µm (Qm=1.6) (n=5), elliptical, brown
(KOH),smooth, without germ-pore (figure e)
Trama: parallel (figure g: white arrow)
Cystidia: not observed
Basidia: (c. 13 x 5.6)µm with 2-4 sterigmata (figure e:red arrow)
Pileipellis: cutis or trichoderm of repent hyphae, some inflated - scalp section
(figure f: black arrow)
Samples: Yungay, Ancash:
*GMA PT 4 PNA 1.8B ,
3846m   9°4'39.629" S 77°39'6.606" W
GMA PT 4 PNA 1.11 ,
3855m   9°4'39.846" S 77°39'6.75" W
GMA PT 5 PNA 1.4BF ,
est.3846m 9°4'39.629" S 77°39'6.606" W
Observations: cf. *Agaricus*: Very similar to those of Zurite and Chinchero,
Cusco that has tentatively been placed in *Micropsalliota* due to burnt amber
brown pore print color - differs only by its odor. Also similar to *A.placomyces*
but without yellow color reaction or phenolic smell.

Section like Xanthodermati or Avenses:

Latently or strongly lutescent (then for Xan. eventually discolouring brownish or vinaceous, but Aven. persistent), annulus intermediate or skirt like, odor phenolic (Xan), KOH positive (Arora 1986:313)

P19   ***Agaricus sp.***   Spores: x̄ (7.6x4.9)μm (Qm=1.5)
      Central Andes: Cerro de Pasco

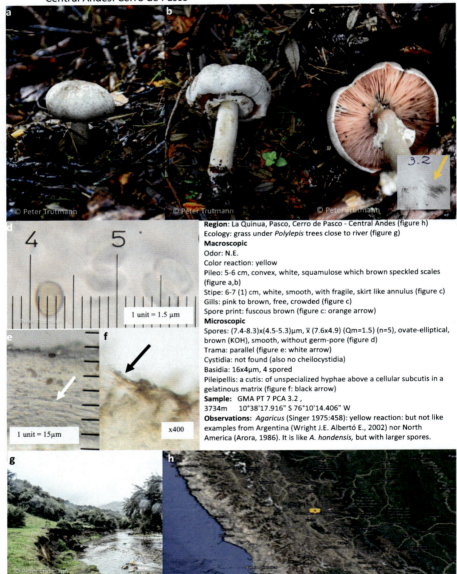

Region: La Quinua, Pasco, Cerro de Pasco - Central Andes (figure h)
Ecology: grass under *Polylepis* trees close to river (figure g)
Macroscopic
Odor: N.E.
Color reaction: yellow
Pileo: 5-6 cm, convex, white, squamulose which brown speckled scales (figure a,b)
Stipe: 6-7 (1) cm, white, smooth, with fragile, skirt like annulus (figure c)
Gills: pink to brown, free, crowded (figure c)
Spore print: fuscous brown (figure c: orange arrow)
Microscopic
Spores: (7.4-8.3)x(4.5-5.3)μm, x̄ (7.6x4.9) (Qm=1.5) (n=5), ovate-elliptical, brown (KOH), smooth, without germ-pore (figure d)
Trama: parallel (figure e: white arrow)
Cystidia: not found (also no cheilocystidia)
Basidia: 16x4μm, 4 spored
Pileipellis: a cutis: of unspecialized hyphae above a cellular subcutis in a gelatinous matrix (figure f: black arrow)
Sample: GMA PT 7 PCA 3.2 ,
3734m   10°38'17.916" S 76°10'14.406" W
Observations: *Agaricus* (Singer 1975:458): yellow reaction: but not like examples from Argentina (Wright J.E. Albertó E., 2002) nor North America (Arora, 1986). It is like *A. hondensis,* but with larger spores.

P20
(1)
Agaricus sp.
Central Andes: Ancash

Spores: x̄ (7.1x4.4)μm (Qm=1.6)

1 unit = 3.75μm

Region: Llanganuco, Yungay, Ancash - Central Andes (figure h)
Ecology: organic material in Polylepis forest (figure h)
Macroscopic
Odor: chemical (phenolic)
Color reaction: none - yellow?
Pileus: 5cm, plane, white with very light cream streaks, smooth
Stipe: 5-6 (1)cm, white to cream, smooth, with intermediate, membranous, fragile type annulus (figures a,b)
Gills: pink to brown, free, crowded (figure c)
Spore print: fuscous brown (figure c: orange arrow)
Microscopic
Spores: (6.8-7.5)x(4.2-4.5)μm, x̄ (7.1x4.4)μm (Qm=1.6) (n=5), elliptical to subglobal, brown (KOH), smooth, without germ-pore (figure e)
Trama: Not evaluated
Cystidia: Not evaluated
Basidia: Not evaluated
Pileipellis: Not evaluated
Sample:
　　　GMA PT 5 PNA 1.15A
3850m　9°4'43.187" S 77°39'8.376" W
Observations: *Agaricus* (Singer 1975:458): uncertain if yellow staining

Section specimens with uncertain properties:
rufescent or not discolouring, with phenolic or sharp odor, and intermediate to skirt like annulus

P21   ***Agaricus sp***   Spores: x̄ (6.7x4,4)µm (Qm=1.5)
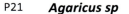
Southern Andes: Cusco

Region: Zurite, Anta, Cusco (figure e)
Ecology: organic matter under *Polylepis* forest (figure d)
Macroscopic
Odor: chemical (phenolic), taste: spicy
Color reaction: none to slightly red.
Pileus: 5-11cm, plane (mature), white, smooth (figure a)
Stipe: 5-7(1-1.5)cm, white, bulbous, smooth with skirt like annulus (figures a,b)
Gills: pink to brown, free, crowded (figure b)
Spore print: fuscous brown (figure b: orange arrow)
Microscopic
Spores: (6.0x7.5)x(4.1x4-5)µm, x̄ (6.7x4,4)µm (Qm=1.5) (n=7), elliptical, brown (KOH), smooth, without germ-pore (figure c)
Cystidia: none observed
Basidia: short, c.(26)x(3.7-7.5)µm, 2-4 spores
Pileipellis: not evaluated
Sample:
GMA PT 2 PSA 6.5
3902 m   13°25'54.876" S   72°15'18.263" W
Observation: *Agaricus* (Singer 1975:458); characterized by a smooth pileus.

1 unit = 3.75 µm

P22 (2) *Agaricus sp.*
Northern Andes: Cajamarca

Spores: (x̄ (5.6x3.7)μm (Qm=1.5)

Region: Cumbemayo, Cajamarca, Cajamarca - Northern Andes (figure e)
Ecology: soil, Ichu based native grassland (figure e)
Macroscopic
Odor: sharp - spicy champignon
Color reaction: none
Pileus: 4-7 cm, plane, white with brown steaks with darker center (figure a)
Stipe: 5-7(1-1.5) cm, white, smooth, bulbous, with membranous, skirt type annulus (figures a, b)
Gills: brown, free, crowded (figure c)
Spore print: fuscous brown (figure c: orange arrow)
Microscopic
Spores: (4.5-6.4)x(3.7)μm, x̄ (5.6x3.7)μm (Qm=1.5) (n=5) elliptical, smooth, brown (KOH), without pore (figure d)
Trama: not evaluated
Cystidia: not evaluated
Basidia: not evaluated
Pileipellis: not evaluated
Samples: *GMA PT 3 PNA 4.7
3638 m   7°11'18.09" S 78°34'36.611" W
GMA PT 3 PNA 4.16
3585 m   7°11'30.468" S 78°34'43.583" W
Observations: *Agaricus* (Singer 1975:458): Very similar to specimen (GMA PT 3 PSA 7.21) found in Cusco except spore print color fuscous rather than burnt amber brown.

1 unit = 3.75μm

P23   *Agaricus sp.*   Spores: x̄ (4.9x3.9) µm (Qm=1.2)
Northern Andes: Tumbes

d

1 unit = 3.75µm

Region: Tumbes, Tumbes
Ecology: organic matter, leaf litter in north-western Andean mountaine forest
Macroscopic
Odor: chemical (phenolic)
Color reaction: none
Pileus: 6-11cm, convex-plane with edges of an even border of veil membrane, white with fine fiber-like light brown scales with darker center (fig.a,b)
Stipe: 5-8 (0.5-1)cm, white, smooth, with fragile membranous skirt type annulus (figure c)
Gills: white (only one young specimen), free, crowded (figure c)
Spore print: fuscous to purple-brown (figure c: orange arrow)
Micrsocopic
Spores: elliptical-subglobose, (4.1-5.6)x(3.8-4.1) µm, x̄ (4.9x3.9) µm (Qm=1.2) (n=6), smooth, brown (KOH), without pore
Trama: not evaluated
Cystidia: not found
Basidia: not found
Pileipellis: a cutis : mycelium
Sample: GMA PT 5 PNA 6.17
136 m   3°46'35.777" S 80°20'51.474" W
Observation: *Agaricus* (Singer 1975:458). sample in poor condition.

e

Coprinus

Coprinus comatus   Spores: x̄ (14x8), (Qm= 1.7)
Southern Andes: Ayacucho, Cusco, Puno

Region: Ayacucho, Cusco, Puno - Southern Andes (figure f)
Ecology: organic matter, agro-pastures, compost, puna (figure f)
Macroscopic
Odor: mildly chlorinated but agreeable
Pileus: 2-3x10-cm, broadly parabolic to campanulate, white with prominent scales, turning dark beginning at outer edge at maturity, deliquescent (figures a, b, c)
Stipe: 10(0.5-1)cm, white, smooth, hollow, fibrous without annulus (figures a,b,c,d)
Gills: pink to black, free, crowded, deliquescent when mature (figure d)
Spore print: black
Microscopic
Spores: (11-17)x(7-9)μm, x̄ (14x8), (Qm= 1.7) (n=5), dark brown to black, elliptical, smooth, without germ-pore (figure e)
Samples:
Huancane, Puno GMA PT 1 PSA 10.3:
3830 m   15°10'46" S 69°50'38.62" W
                    GMA PT 3 PSA 8A.4,
3846 m   15°10'46" S 69°50'38.62" W
Cusco, Cusco: GMA PT 2 PSA 5.13,
3556 m   13°34'21.696" S 71°51'25.158" W
                    GMA PT 2 PSA 4.11F
Puquio, Ayacucho: GMA PT 2 PSA 1.4 F
3196 m   :14,41.1806S 74,07.0176W
Anta, Cusco: GMA PT 3 PSA 6.1,
3846 m   15°10'46" S 69°50'38.62" W
Observations: *Coprinus comatus* (O.F.Müll.) Pers. with spores (10-16 x(7-9)μm (Arora, 1986: 345). Traditional Quechua:'C'hopca'(Holguin de, 1608)

1 unit = 3.75 μm

Hymenagaricus

P25 **Agaricaceae:** *cf.* ***Hymenagaricus*** Spores: x̄ (6.3x4.4)µm (Qm=1.4)
  Central Andes: Lima

Region: Zarate, Huarochiri, Lima - Central Andes (figure j)
Ecology: in leaf litter and organic material below *Polylepis* forest (figure i)
Macroscopic
Odor: mild a little peppery
Color change: red? (figure c: red arrow)
Pileus: (7-9)cm, plane to slightly depressed, squamulose with small fragile red-brown scales covering a white base, (figure c)
Stipe: (4-20)cm, strikingly long, white, farinose and with white scales on lower side, with fragile membranous intermediate type annulus (figures a,b)
Gills: reddish brown, free, crowded (figure d)
Spore print: appears fuscous brown (sample putrefied) (fig. d: orange arrow)
Microscopic
Spores: (6.0-7.5)x(4.4-4.5)µm, x̄ (6.3x4.4)µm (Qm=1.4) (n=5), oval-elliptical, smooth, brown (KOH) without germ-pore (figure e)
Trama: parallel, slightly dextrose? (figure f: white arrow)
Cystidia: not found
Basidia: c.(7.5x2)µm, with 4 spores, slightly dextrose?
Pileipellis: a dermis: appears hymeniform with parenchymal like cells above a mycelial subcutis - assuming supracutis is not remnant veil tissue with globular cells - radial sections (figures g,h: black arrows)
Sample: GMA PT 6 PCA 1.12 ,
2948 m   11°55'53.729" S 76°28'53.219" W
Observations: Agaricaceae: cf. *Hymenagaricus by its* hymeniform like pileipellis (if cells part of veil then an *Agaricus* due to parallel mycelial subcutis) , brown spores, carpophore form and small spore size and form (see (Reid and Eicker, 1995). The Genus is apparently widespread in the tropics. This specimen is characterized its strikingly elongated stipe.

1 unit = 1.5 µm

1 unit = 3.75 µm

1 unit = 1.5 µm

Agaricaceae: *cf. Hymenagaricus*   Spores: x̄ (6.1x4.4) µm (Qm=1.4)
Central Andes: Lima

Region: Zarate, Huarochiri, Lima - Central Andes (figure i)
Ecology: leaf litter and organic matter in *Polylepis* forest (figure h)
Macroscopic
Odor: mild (and conspicuously consumed by insects)
Color reaction: red
Pileus: (3-4)cm, lepiotoid to plane to slightly uplifted when mature, squamous with large broad red brown scales (red arrow) , over a white surface, center darker colored (figures a,b)
Stipe: (7-10)cm, white, smooth, with reddish streaks, and with intermediate to skirt like annulus (figures b,c)
Gills: pink to brown, free, crowded (figure c)
Spore print: fuscous brown? (sample putrefied) (figure c:orange arrow)
Microscopic
Spores: (6.0-6.3)x(4.2-4.5)µm, x̄ (6.1x4.4)µm (Qm=1.4) (n=5), oval-broadly elliptical, smooth, brown (KOH) without germ-pore (figure d)
Trama: parallel, pseudoamyloid (figure e: white arrow)
Cystidia: not found
Basidia: short, pseudoamyloid, with 2 or more spores
Pileipellis: a dermis: hymeniform in a gelatinous matrix (no mycelium visible). Cells in supra and subpellis look bricklike - radial (figure f: black arrow) and scalp sections(fig.g:red arrows)
Sample: GMA PT 6 PCA 1.15A ,
295 m   11°55'54.372" S 76°28'53.843" W
Observations: Agaricaceae: (not *Agaricus*) tentatively *Hymenagaricus*: due to hymeniform pileipellis, spores small poreless, brown. Tropical Genus reported mainly from Africa and Asia (Reid and Eicker, 1995)

Melanophyllum

cf. Melanophyllum sp.   Spores: x̄ (4.7x3.8)μm (Qm=1.2)
      Central Andes: Ancash

Region: Recuay, Ancash - Central Andes (figure h)
Ecology: organic material below Acacias stand close to river (figure g)
Macroscopic
Odor: aromatic (fresh) like 'bouillon' (dry)
Color reaction: None
Pileus:1.5-3 cm, umbonate-plane, lepiotoid, fragile, white with fine dark pinkish coloration and scales above white base and a darker umbo, with shaggy margin (figure a)
Stipe: 3-6 cm, thin, white, with layer of large, long white pealing strips membrane (epidermis or epithelium?), traces of an annulus (figures d,c)
Gills: brown, free, close to crowded (figure c)
Spore print: greyish-olive to brownish purple (figure c: orange arrow)
Microscopic
Spores: small, (4.5-5.6) x (3.5-3.8)μm, x̄ (4.7x3.8)μm (Qm=1.2) (n=5), subglobose, olive brown with vacuoles (KOH), smooth, without germ-pores (figure d)
Trama: parallel (figure f: white arrow)
Cystidia: not observed
Basidia: with 4 spores, (figure e: black arrow)
Pileipellis: not evaluated
Sample: GMA PT 3 PNA 9.20,
1661 m 10°9'20.309" S 77°31'19.475" W
Observations: *Melanophyllum* (Singer 1975:458): by spore print color. Small subglobose spores, shaggy pileus, and pealing strips of membrane characterize this species

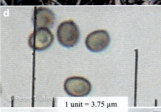

1 unit = 3.75 μm

Micropsalliota

P28   *cf.* **Micropsalliota sp.**   Spores: x̄ (6.6x3.8)μm (Qm=1.8)
Southern Andes: Cusco

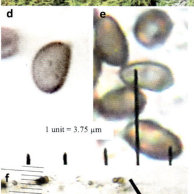

1 unit = 3.75 μm

Region: Zurite, Anta y Chinchero, Cusco - Southern Andes (figure h)
Ecology: organic matter in *Polylepis* forest and Eucalyptus stands (figure g)
Macroscopic
Odor: chemical (Phenolic) but mild
Color reaction: none or mildly red
Pileus: 5-15cm, convex (young) to plane (mature), white with dark brown scales and dark center, or after rain white with brown/red colored rays (figures a,b)
Stipe: 7-15 (1-1.5)cm, white, smooth, bulbous, with white skirt type membranous annulus (figure c)
Gills: pink to brown, free, crowded (figure c)
Spore print: burnt amber brown (figure c: orange arrow)
Microscopic
Spores: (5.6-7.5)x(3.4-4.1)μm, x̄ (6.6x3.7μm) (Qm=1.8) (n=5), elliptical, brown (KOH), smooth, without germ-pore (figures d,e)
Trama: parallel or interwoven (figure f: black arrow))
Basidia: c.(18-22)x(4)μm with 2-4 spores
Samples:
Zurite, Anta, Cusco: GMA PT 3 PSA 7.21, 4012 m   13°25'47.1" S 72°15'24.401" W
Zurite, Anta, Cusco: GMA PT 3 PSA 7.25, 3819 m,   13°23'27.414" S 72°2'32.237" W
Chinchero, Cusco: GMA PT 3 PSA 3.13, 3819 m,   13°23'27.414" S 72°2'32.237" W
Observations: keys to *Micropsalliota* (Singer 1975:458): by burnt amber brown print, smooth spores, pseudoamyloid reaction and carpophore form .

cf. *Micropsalliota sp.*
Spores: (x̄ (5.7x3.9) µm (Qm=1.5)
Northern Coastal Hills: Tumbes

Region: Tumbes, Tumbes - Northern Andes (figure h)
Ecology: soil, grassland with native trees (figure h)
Macroscopic
Odor: mild and like an edible *Agaricus* (fresco)
Color reaction: no.
Pileus: 4-8cm, broadly convex to plane, white with a grey center (calotte), smooth (figures and and b) with remnant scales of the vale when young (figure c)
Stipe: 3-10 (0.5-1)cm, white, smooth, with an intermediate type membranous annulus (figure b and d)
Gills: dark brown, free, crowded (figure d)
Spore print: burnt amber brown (figure d: orange arrow)
Microscopic
Spores: (4.5-6.4)x(3.4-4.2)µm, x̄ (5.7x3.9) µm (Qm=1.5) (n=12), elliptical to subglobose, smooth, dark brown (KOH), without germ-pore (figure e)
Trama: not evaluated
Cystidia: not observed
Basidia: c.(11.3x7.5) µm, with 2 to 4 spores (figure f: red arrow)
Pileipellis: a cutis: of interwoven unspecialized hyphae (scalp section) (figure g: black arrow)
Samples: GMA PT 5 PNA 5.11
67 m   3°44'29.309" S 80°24'32.813" W
Tumbes, Tumbes: GMA PT 5 PNA 5.3
61 m   3°45'23.111" S 80°23'9.515" W
Observationes: *Micropsalliota* (Singer 1975: 458). Not *Agaricus* due to burnt amber (not fuscous) spore color. Eaten by maggots.

1 unit = 3.75 µm

Agaricus

P30 **_Agaricus_ cf. _aridicola_**
Southern Andes: Arequipa

Spores: x̄ (5.9x5.9)μm (Qm=1.0)

Region: Arequipa, Arequipa - Southern Andes (figure g)
Ecology: soil in agro-pasture (figure f)
Macroscopic
Color reaction: not evaluated
Pileus: (2-3)cm, broadly convex white-cream, striated and zoned otherwise smooth, dull and dry (figure a).
Stipe: (3-4)cm, white-cream, smooth, without annulus (figure a)
Gill-like gleba: green-brown, like compacted, fused ramified plates (white arrow), that apear free of the stipe (figure b:white arrow).
Spore print: brown (figure b:orange arrow)
Microscopic
Spores: (5.5-6)x(5.5-6)μm, x̄ (5.9x5.9)μm (Qm=1.0) (n=5), globose, smooth, thick walled, without germ-pore, hilar appendage (pedicel) short (figure c)
Basidia: pigmented, with 4 sterigmata (figure d: red arrow)
Pileipellis: a cutis: of repent, inflated, parallel hyphae (figure e: black arrow)
Sample: GMA PT 1 PSA 12.7
3960 m   16,4.5501S 71,22.8747W
Observations: *Agaricus:* conforms to the gastroid *Agaricus aridicola* (syn. *Longula*) with smooth, globose, thick-walled spore (5.0-6.2)x(4.6-5.7)μm with a pileicutis. The alternative, *A. deserticola*, has lightly ornamented spores (5.5-7.0)x (5.0-6.0)μm and size and is shaped differently.

Bovista

P31   *Bovista sp.*   *('el cerebro')*   Spores: (x̄ (5.8x5.8)μm (Qm=1.0)
(12)   Southern Andes: Arequipa, Ayacucho, Puno

Peter Trutmann

Region: Arequipa, Puno, Ayacucho, Cusco - Southern Andes (figure f)
Ecology: soil, eroded land, tracks, degraded, overgrazed pasture and puna grasslands (figure f)
Macroscopic
Fruiting body: c. 2-3x2x2cm, globose puffball, white exoperidium with large wart-like mosaic surface (reminiscent of a brain), with rudimentary or absent sterile base, rupturing without apical pore (figures a,b,c)
Spore mass: brown (figure c)
Microscopic
Spores: globose, smooth, (5.0-6.5) x (5.0-6.5)μm, x̄ (5.8x5.8)μm (Qm=1.0) (n=5) green-brown (KOH) with short pointed pedicel of 2-13 μm(figures d,e)
Capillitium: very sparsely branched
Peridium layers: not evaluated
Samples:
Ilave, Puno: GMA PT 1 PSA 8.2
est. 3867 m   16°5'51.51" S 69°37'8.316" W
Huancane -Putina, Puno
GMA PT 3 PSA 8A.8 F
GMA PT 3 PSA 8B.5F,
GMA PT 2 PSA 9.5
3,827.0 m   15°10'46" S 69°50'38.62" W
Lampas, Puno
GMA PT 1 PSA 0.3
4389 m   15°43'51.503" S 70°50'57.215" W
Megar, Puno: GMA PT 1 PSA 7.4BF
4219m GPS: 14°29'44.868" S 70°56'49.446" W
Arequipa
GMA PT 3 PSA 8.4
3920 m   16°3'25.685" S 71°21'19.986" W;
GMA PT 1 PSA 12.8B,
3960 m   16°4'33.006" S 71°22'52.482" W;
GMA PT 2 PSA 10.5
3,960 GPS: 16°4'33.018" S 71°22'52.493" W,
GMA PT 3 PSA 8.9F
3,827 m   15°10'46" S 69°50'38.62" W
Lucanas, Ayacucho:
GMA PT 2 PSA 2.6B ,
GMA PT 2 PSA 2.4BF
4407m   14°37'59.214" S 73°57'26.369" W
Observations: *Bovista* (Arora 1989:696): larger spores than similar Ancash samples. (P32)

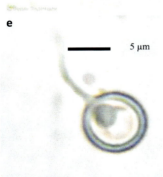

5 μm

5 μm

P32
(3)
Bovista sp.

Central Andes: Ancash

Spores x̄ (5.4x5.4)μm, (Qm=1.0)

Region: Recuay, Ancash - Central Andes (figure f)
Ecology: soil, often overgrazed puna grasslands (figure f)
Macroscopic
Fruiting body: (1-2x1)cm, exoperidium white, oval, ornamented with wartlike mosaic resembling a brain, without apical pore, with rudimentary to absent sterile base (figures a,b).
Spore mass: brown
Microscopic
Spores: (4.8-5.6)μm, x̄ (5.4x5.4)μm, (Qm=1.0) (n=5), globose to , smooth, green-brown (KOH), infrequently with pointed pedicel (7.5-15)μm, in pockets (figure c, d)
Capillitium: non branched hyphae (figure e: black arrow)
Samples:
GMA PT 3 PNA 9.2
3745 m   9°51'25.667" S 77°24'29.592" W
GMA PT 3 PNA 9.5
3875 m   9°56'34.931" S 77°22'55.619" W
GMA PT 3 PNA 9.6
3873 m   9°56'34.541" S 77°22'56.616" W
Observations: *Bovista* (Arora 1986:696): smaller spored 'el celebro' Bovista (P31)

1 unit = 3.75μm

© Peter Trutmann

P33 **_Bovista sp_**
Northern Andes: Cajamarca

Spores: (x̄ (4.9x4.9)μm c.(Qm=1.0)

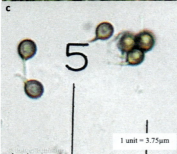

Region: Cajamarca, Cajamarca - Northern Andes (figure e)
Ecology: soil in puna grasslands (figure d)
Macroscopic
Fruiting body: c. (4x3.5x3)cm, exoperidium white ornamented with mosaic of small warts, and tiny farinose spikes, without sterile base without apical opening (figures a,b)
Spore mass: white (only young samples)
Microscopic
Spores: globose to subglobose (4.0-5.6)x(4.0-5.6)μm, x̄ (4.9x4.9)μm (Qm=1.0), (n=5), green-brown (KOH), smooth, with short pedicel (3.7-4.5)μm (figure c)
Capillitium: not evaluated
Peridium layers: not evaluated
Sample: GMA PT 3 PNA 4.17
3557 m   7°11'30.089" S 78°34'42.984" W
Observations: _Bovista_ (Arora 1986:696): characterized by fine warts.

1 unit = 3.75μm

Bovista sp.

Photographic record only

Southern Andes: Cusco

Region: Anta, Cusco

Ecology: soil in degraded shrubbed grassland

Macroscopic

Fruiting body: (2-3x2x2)cm, globose, white exoperidium , reticulated with small darker colored warts, with minimal or absent sterile base, without apical pore spitting to release spores (figures a,b)

Peridium: two walled

Spore mass: yellow to brown (figure c)

Microscopic

Spores: not evaluated

Photographic sample only:

GMA PT 3 PSA 7.30CF.

3985 m   13°25'46.494" S 72°15'19.721" W

Observations: *Bovista* (Arora 1986:696): characterized by darker warts than P33.

Bovista sp.   Not evaluated microscopically
SOUTHERN ANDES: CUSCO

© Peter Trutmann

Region: Quispicanchi, Cusco -Southern Andes (figure c)
Ecology: compacted soil on a walking track in shrubbed grassland by Inka ruins (figure c)
Macroscopic
Fruiting body: c.(2-3x2x2)cm, white exoperidium without conscious ornamentation, without sterile base, splitting to release spore mass (figure a)
Peridium: two walled (figure b)
Spore mass: brown (figure b)
Microscopic
Spores: not evaluated
Sample: GMA PT 3 PSA 5.13B
3,514 m   13°34'9.558" S 71°47'3.665" W
Observation: *Bovista* (Arora 1986:696): characterized by a non-ornamented, smooth, dull white peridium

Quispicanchi, Cuzco. Peru

Bovista sp.　　　Spores: x̄ (4.5x4.3)μm (Qm=1.0)

Central Andes: LIma

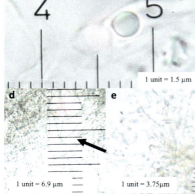

1 unit = 1.5 μm

1 unit = 6.9 μm　　1 unit = 3.75μm

Region: Canta, Lima (- Central Andes figure f)
Ecology: degraded environments such as shrubbed grassland
(figure f)
Macroscopic
Fruiting body: (1x2x1)cm, exoperidium cream, irregularly
shaped, smooth, with minimal or absent sterile base (figures
a,b)
Spore mass: white (immature)
Microscopic
Spores: (4.5-4.8)x(3.7-4.5)μm, x̄ (4.5x4.3)μm (Qm=1.0) (n=5),
globose a subglobose, hyaline (KOH), smooth, without pedicel
(figure c)
Capillitium: hyaline, very sparsely branched (figure d: black
arrow)
Basidia: with 4 sterigmata (figure e: white arrow)
Sample: GMA PT 7 PCA 1.19
3594 m   11°26'29.7" S 76°35'7.368" W
Observations: *Bovista* (Arora 1986:696): differentiated by
irregular white, carpophore shape and spores.

P37 **_Bovista sp._**
Central Andes: Junín

Spores: x̄ (7.4x7.1)µm (Qm=1.0)

© Peter Trutmann

1 unit = 1.5 µm

1 unit = 6.9 µm

Region: Yauli, Junín - Central Andes (figure g)
Ecology: in puna grasslands (figure f)
Macroscopic
Fruiting body: c. (1x2x1)cm, oval, with smooth, white exoperidium and minimal sterile base (figures a,b)
Peridium walls: not evaluated
Spore mass: yellow-brown (figure c: orange arrow)
Microscopic
Spores: (6.7-8.2)x(6.7-7.5)µm, x̄ (7.4x7.1)µm (Qm=1.0) (n=5), globose to subglobose, hyaline to green (KOH), smooth, without pedicel (figure d).
Capillitium: hyaline, not or rarely branching (figure e: black arrow)
Basidia: not found
Sample: GMA PT 7 PCA 2.2
4604 m   11°16'53.568" S 76°26'46.398" W
Observations: _Bovista_ (Arora 1986:696): differentiated by altitude and location, smooth, white carpophore shape, size, and spore mass color.

© Peter Trutmann

THE MACROFUNGI OF ANDEAN PERU Part 1

P38 *Bovista sp*
Northern Andes: Cajamarca

Spores: x̄ (4.2x4.2)μm c.(Q=1.0)

Region: Cerro el Castillo, Cajamarca - Northern Andes (figure e)
Ecology: soil in agro-pastures by lago Suyacocha (figure d)
Macroscopic
Fruiting body: c. (1-3.5)x(1-2)cm, white exoperidium with large wart-like mosaic ornamentation, without sterile base (figures a,b)
Peridium walls: not evaluated
Spore mass: brown
Microscopic
Spores: (4.0-4.5)x (4.0-4.5), x̄ (4.2x4.2)μm (Q=1.0) (n=5), globose to subglobose, green-brown (KOH), smooth, with very short pedicel (1-4)μm (figure c)
Capillitium: not found
Sample: GMA PT 3 PNA 5.3
2986 m   7°11'28.644" S 78°22'20.519" W
Observations: *Bovista* (Arora 1986:696): differentiated by white, irregular carpophore form, location and spore size.

1 unit = 3.75μm

Bovista sp.
Central Andes: Ancash

Spores: x̄ (4.5)µm c.(Qm=1)

© Peter Trutmann

© Peter Trutmann

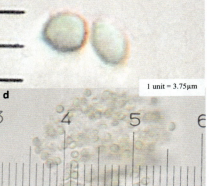

1 unit = 3.75µm

Region: Yungay, Ancash - Central Andes (figure e)
Ecology: highland pastures connected to *Polylepis* forest (figure e)
Macroscopic
Fruiting body: (1x2x1)cm, globose puffball, exoperidium white with pimple-like protrusions, and without, or with minimal, sterile base, breaking at the apex at maturity (red arrow) to release spores (figures a,b)
Peridium wall: not evaluated
Spore mass: yellow green
Microscopic
Spores: (4.0-5.6)µm, x̄ (4.5)µm c.(Qm=1) (n=5), globose to oval, smooth, hyaline (KOH), without pedicel (figure c)
Capillitium: no hyphae visible at maturity (digested?)(figure d)
Sample: GMA PT 3 PNA 8.18B
3806 m 9 °4'49.337" S 77°39'20.064" W
Observations: *Bovista* (Arora 1986:696): differentiated by location, carpophore form and spore size and form.

© Peter Trutmann

Bovista sp. Photographic record only
Central Andes: Ancash

Region: Yungay, Ancash - Central Andes (figure d)
Ecology: soil close to the shores of lake Llanganuco (figure d)
Macroscopic
Fruiting body: c. (1x1)cm, spherical to oval shaped, exoperidium
white, ornamented with small white spines, without sterile
base, without a classic porous opening, breaking open when
mature (figures a,b,c)
Spore mass: brown (figure c: red arrow)
Photographic sample only:
GMA PT 3 PCA 1.15F
3800m,   9°4'40.847" S 77°39'7.89" W
Observations: *Bovista* (Arora 1986:696): differentiated by
location, carpophore size, small white spines, brown spore
mass and characteristic splitting of the peridium.

Bovista sp   Spores: x̄ (4.0x4.0)μm c.(Q=1.0)
Northern Andes: Cajamarca

Region: Cajamarca, Cajamarca (figure d)
Ecology: soil in agro-pasture (figure d)
Macroscopic
Fruiting body: c.(2-2.5)x (1.5-2)cm, exoperidium white ornamented
with fine spines, with sterile base containing fine rhizome like
structures (figures a,b)
Spore mass: yellow-green
Microscopic
Spores: (3.7-4.5)x(3.7-4.5)μm, x̄ (4.0x4.0)μm (Q=1.0) (n=5), globose
to subglobose, green (KOH), smooth, ornamented with spines,
without pedicel (figure c: black arrow)
Capillitium: disintegrated
Sample: GMA PT 3 PNA 5.10
2748 m   7°13'30.72" S 78°16'35.028" W
Observations: _Bovista_ (Arora 1986:696): differentiated by white
carpophore size with small spines, spore size and shape, and
location.

Bovista sp.   Spores: x̄(4.6x4.5)μm (Q=1.0)
Central Andes, Cerro de Pasco

d

5

                                                    5 μm

e

1 unit = 10μm

Region: La Quinoa, Cerro de Pasco - Central Andes (figure f)
Ecology: soil in grass by *Polylepis* forest and river (figure f)
Macroscopic
Fruiting body: (1x2 x1)cm, oval shaped, exoperidium white to
cream, ornamented with very small warts like toad skin, with
small sterile base - not metallic colored when mature (figures
a,b)
Spore mass: brown
Microscopic
Spores: (4.5-6.0)x(3.8-4.5)μm, x̄(4.6x4.5)μm (Qm=1.0) (n=5)
oval with thick wall, green with yellow wall (KOH) without
pedicel
Capillitium: spores produced in pockets possibly with
membranes. Hyphae branched pigmented apparently
disintegrating at maturity (figure e: black arrow)
Sample: GMA PT 7 PCA 4.1B
3270 m   10°38'10.728" S 76°10'17.592" W
Observations: *Bovista* (Arora 1986:696): differentiated by
carpophore form, its unusual thick-walled oval spores and
location.

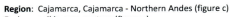

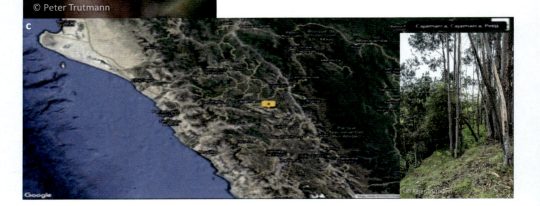

Region: Cajamarca, Cajamarca - Northern Andes (figure c)
Ecology: soil in agro-pasture (figure c)
Macroscopic
Fruiting body: (2-3x2)cm, irregular shape, white, with thin exoperidium commonly decorated with a mosaic of darker colored small warts, minimal sterile base with rhizome like mycelium (figure a,b).
Spore mass: not assessed
Microscopic
Not evaluated
Sample: GMA PT 3 PNA 5.14
2718 m   7°13'59.364" S 78°16'5.688" W
Observations: _Bovista_ (Arora 1986:696): differentiated by the combination of its dark wartlike mosaic exoperidium and irregular shape. Similar mosaic form to P34F from Cusco and by its irregular form with P38 also from Cajamarca. The sample requires microscopic evaluation

© Peter Trutmann

Bovista sp   Photographic record only
Northern Andes: Piura

Region: Huancabamba, Piura - Northern Andes (fig. e)
Ecology: soil on the banks of a dry river (figure e)
Macroscopic
Fruiting body: c. (2x3x2)cm, spherical to oval,
exoperidium white changing to metallic blue-grey,
ornamented with small spines, without a sterile base
Spore mass: not able to determine
Photographic sample only:
GMA PT 6 PNA 1.1F
2158m   5°50'26.147" S 79°30'20.238" W
Observations: *Bovista* (Arora 1986:696):
differentiated by carpophore form with farinose
surface and change from white to metallic color, as
well as by location.

Bovista sp.   Spores: x̄ (4.7x4.3) (Qm=1.1)

Central Andes: Cerro de Pasco

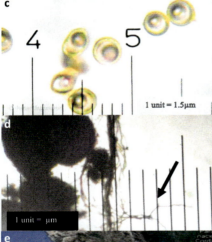

Region: La Quinoa, Cerro de Pasco - Central Andes (figure e)
Ecology: pasture associated with a *Polylepis* forest (figure e)
Macroscopic
Fruiting body: c. (1x2-3x1)cm, oval shaped, thin exoperidium white, with darker warts, turning metallic black when mature, releasing spores through an uneven apical opening (pore?), with a minimal sterile base (figure a,b)
Spore mass: brown
Microscopic
Spores: (4.5-5.3)x(3.8-4.8)µm, x̄ (4.7x4.3)µm (Qm=1.1) (n=5), globose to subglobose, smooth, yellow (KOH) with brown colored vacuole, infrequently with short pedicel (figure c)
Capillitium: pigmented, of frequently branched hyphae, without membrane (figure d: black arrow)
Sample: GMA PT 7 PCA 4.1
3062 m   10°38'8.633" S 76°10'18.174" W
Observations: *Bovista* (Arora 1986:696): differentiated by carpophore form and change of color, spore size and shape as well as location.

THE MACROFUNGI OF ANDEAN PERU   Part 1

P46
(2)

cf. *Bovista sp.*
Central Andes: Junín

Spores:, x̄ (7.3x6.7)µm (Qm=1.1)

a

b

© Peter Trutmann

© Peter Trutmann

c

4   5

1 unit = 1.5 µm

d

4   5   6

1 unit = µm

Region: Yauli, Junín - Central Andes (figure e)
Ecology: soil in Ichu, puna grasslands (figure e)
Macroscopic
Fruiting body: c. (1x2x1)cm, oval shaped, exoperidium white, smooth, white when young turning metallic black when mature, splitting apically when mature, with minimal sterile base
Spore mass: brown
Microscopic
Spores: (6.0-8.0)x(6.0-7.2)µm, x̄ (7.3x6.7)µm (Qm=1.1) (n=5) globose to subglobose, dark brown (KOH) with short spines (0.8 µm) without pedicel (figure c)
Capillitium: pigmented, infrequently branched (figure d: black arrow)
Basidia: 4 spored
Sample:
GMA PT 7 PCA 2.3
GMA PT 7 PCA 2.5,
4607 m   11°16'53.568" S 76°26'46.404" W
Observations: *Bovista* (Arora 1986:696): placed provisionally as *Bovista*. Differentiated by metallic black color when mature, spore mass color, spore form and size, as well as location.

e

Peru

Yauli, Junín, Peru

© Peter Trutmann

P47 ***Bovista sp.***   Spores: x̄ (4.6x4.6)µm (Qm=1.0)
(2)   Central Andes: Ancash

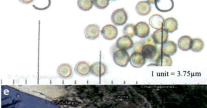

Region: Yungay, Ancash - Central Andes (figure e)
Ecology: soil and organic matter in *Polylepis* forest (figure e)
Macroscopic
Fruiting body: (1-3x1-3)cm, sphere to disc shaped, exoperidium metallic brown, with a sine mosaic pattern, with apical pore, without sterile base (figures a,b)
Spore mass: brown (figures a,c)
Microscopic
Spores: (4.5-4.8)x(4.5-4.8)µm, x̄ (4.6x4.6) (Qm=1), (n=5), globose, smooth to finely ornamented, green (KOH), without pedicel (figure d)
Capillitium: not found
Sample:
GMA PT 3 PNA 8.23B
3840m,   9°4'40.847" S 77°39'7.89" W
GMA PT 5 PNA 1.6CF.
3844 m   9°4'44.496" S 77°39'9.413" W
Observation: *Bovista* (Arora 1986:696): differentiated by carpophore form and color, as well as spore size and shape, spore mass color and location.

P48 *Bovista sp.*
Central Andes: Ancash

Spores: x̄ (4.7x5.4)μm (Qm=1.1),

Region: Yungay, Ancash - Central Andes (figure g)
Ecology: soil, in *Polylepis* forest (figure g)
Macroscopic
Fruiting body: (2.5-3) to (3.5-4.5)cm, spherical, exoperidium white (figure a) to light brown finely ornamented (figure b), without apical pore, without sterile base (figure c)
Spore mass: light brown to black (figure d)
Microscopic
Spores: (3.4-5.6)x(3.8-6.8)μm, x̄ (4.7x5.4)μm, (Qm=1.1), (n=5) globose to subglobose, smooth, green-brown (KOH), often with pedicel (7.5-15) μm (figure e), spores borne in pockets (figure f: red arrow)
Capillitium: large pigmented, branching hyphae, (figure f: black arrow)
Sample: GMA PT 4 PNA 1.18 3875 m 9°3'53.436" S 77°38'4.253" W
Observations: *Bovista* (Arora 1986:696): differentiated by carpophore form, color and size, lack of apical pore. Similar, to Bo17 (perhaps younger?); however, spore size appears larger and more variable and this sample has pedicels.

1 unit = 3.75μm

1 unit = 15 μm

Bovista sp.   Spores: x̄ (3.9x3.9)µm (Qm=1.0)
Central Andes: Ancash

Region: Bolognesi, Ancash (figure d)
Ecology: soil with moss, under shrubs in pasture (figure d)
Macroscopic
Fruiting body: (2.5x2.5)cm, spherical, exoperidium cream to light brown, ornamented with small spines, without apical pore, and minimal sterile base (figures a,b)
Spore mass: greenish brown
Microscopic
Spores: (3.7-4.5)x(3.7x4.5)µm, x̄ (3.9x3.9)µm, (Qm=1), (n=5) globose a subglobose, smooth to finely ornamented, green (KOH), with very short pedicel (figure c: black arrow)
Capillitium: not examined
Sample: GMA PT 3 PNA 9.8
3813 m   10°9'35.556" S 77°19'39.828" W
Observations: *Bovista* (Arora 1986:696): similar carpophore ornamentation to P48, but separated by spore shape and size from both P48 and P47.

1 unit = 3.75µm

Bovista sp.   Spores: x̄ (4.6x4.3)μm (Qm=1.1)
Central Andes: Lima

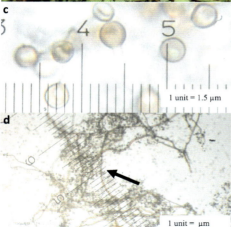

Region: Zarate, Huarochiri, Lima - Central Andes (figure a)
Ecology: moss, beside walking track below shrubs (figs. a,e)
Macroscopic
Fruiting body: (1x2)cm, spherical, exoperidium cream, with small warts and minimal sterile base (figures a,b)
Spore mass: brown
Microscopic
Spores: (3.8-5.3)x(3.8-4.5)μm, x̄ (4.6x4.3)μm (Qm=1.1) (n=5), globose to subglobose, light brown (KOH), smooth, with pedicel (figure c: white arrow)
Capillitium: pigmented, branched (figure d: black arrow)
Sample: GMA PT 6 PCA 1.19B, 2805 m   11°55'45.899" S 76°29'20.91" W
Observations: _Bovista_ (Arora 1986:696): differentiated by carpophore form and size, location, as well as spore size and pedicel.

1 unit = 1.5 μm

1 unit = μm

Bovista sp. Spores: x̄ (4.0x4.0)μm c.(Q=1.0)
Northern Andes: Tumbes

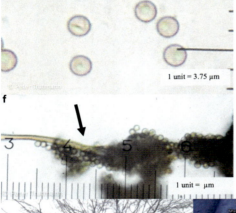

Region: coastal hills, Tumbes - (figure g)
Ecology: sandy soil, in under a native acacia in grassland (figure g)
Macroscopic
Fruiting body: (0.5-1x0.5-1)cm, spherical, exoperidium cream (figures a,b), to yellow when mature (figure d), ornamented when young with small white spinelike warts(figure b), without sterile base (figure c)
Spore mass: yellow (figure d)
Microscopic
Spores: (3.3-4.5)x (3.8-4.5)μm, x̄ (4.0x4.0)μm c.(Qm=1.0) (n=6), globose, light brown with green vacuole (KOH), smooth, without pedicel (figure e)
Capillitium: pigmented, not (or only rarely) branched (black arrow). Spores housed in pockets (figure f)
Sample: GMA PT 5 PNA 5.6
67 m   3°45'23.586" S 80°23'9.018" W
Observations: *Bovista* (Arora 1986:696): differentiated by carphore form and size, its yellow spore mass, together with spore size with minimal pedicels, and location.

1 unit = 3.75 μm

1 unit = μm

THE MACROFUNGI OF ANDEAN PERU   Part 1

Bovista sp.   Not yet evaluated microscopically
Southern Andes: Cusco

Region: Anta, Cusco - Southern Andes (figure d)
Ecology: soil below shrubs by pasture near *Polylepis* forest (figure d)
Macroscopic
Fruiting body: (4x4)cm, spherical, exoperidium white when young becoming a light colored wartlike surface then mature exposing a dark brown endoperidium, viscous, without sterile base or apical pore (figure a,b)
Spore mass: brown (figure c)
Microscopic
Not evaluated
Sample: GMA PT 2 PSA 6.3C
3843 m   13°26'0.684" S 72°15'14.369" W
Observations: *Bovista* (Arora 1986:696): differentiated by carpophore dark brown color, size and location. Requires further evaluation.

Calvatia

Calvatia cyathiformis   Spores: x̄ (7.1x7.1))µm c.(Qm=1)
Southern Andes: Cusco and Puno

1 unit = 3.75 µm

Region: Puno y Cusco (figure d)
Ecology: agro-pastures (figure d)
Macroscopic
Fruiting body: (8x6x4)cm, exoperidium white turning brown and
blue-gray with age, smooth, without apical pore, splitting open to
release spore mass, with minimal sterile base (figure a)
Spore mass: firm white then powdery blue (figure b)
Microscopic
Spores: (5.6-8)µm, x̄ (7.0)µm (Qm=1) (n=13), globose, blue -green
brown (KOH), with short spines, and no pedicel (figure c)
Samples:
Sullistani, Puno   GMA PT 1 PSA 9.11
3915m :   15°43'17.388" S 70°9'31.896" W
Huanco-Lima, Puno GMA PT 3 PSA 8A.2
                   GMA PT 1 PSA 10.6
3838m   15°11'13" S 69°51'19" W
                   GMA PT 3 PSA 8A.1
3827m   15°10'46" S 69°50'38.62" W
Lampa, Puno   GMA PT 1 PSA 1.3:,
3849 m   15°18'46.56" S 70°12'42.528" W
Putina, Puno   GMA PT 3 PSA 8B.1F 3
3977m   14°54'40.219" S 69°52'28.079" W
Chinchero, Cusco GMA PT 3 PSA 3.22B
3749 m   13°23'15" S 72°2'25" W
Observations: *Calvatia* (Arora 1986:634): *C. cyathiformis (*Bosc.)
Morgan: differentiated by its blue spore mass *and bluish-brown
spores between 3.5-7.5 µm*

P54 *Calvatia sp.*   Spores: x̄ (9.7)μm (Qm=1)
Southern Andes: Cusco

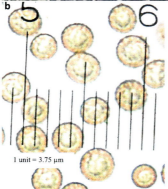

1 unit = 3.75 μm

Region: Zurite, Anta, Cusco - Southern Andes (figure c)
Ecology: grassland with native trees (figure c)
Macroscopic
Odor: putrefying smell when old
Fruiting body: (6-7 x 4)cm, metallic purple grey when mature, smooth, without sterile base, splitting to release spore mass (figure a)
Spore mass: brown-purple
Microscopic
Spores: (7.0—11.2)μm, x̄ (9.7)μm (Qm=1) (n=5), spherical, with spines, greenish brown (KOH) with vacuoles, without pedicel (fig. b)
Capillitium: No structures observed (sample very mature)
Sample: GMA PT 3 PSA 7.32
c 3670 m   13°26'12.048" S 72°15'10.212" W
Observations *Calvatia* (Arora 1986:634**)**: similar carpophore and spore color to *C. cyathiformis*, but differentiated by substantially larger spore size (c.f. *C. cyathiformis* (3.5-7.5)μm)

P55F *Calvatia cf. pachydermica.*   Photographic record only
Southern Andes: Cusco

© Peter Trutmann

© Peter Trutmann

Region: Cusco, Cusco (figure d)
Ecology: soil in highland pastureland (figure d)
Macroscopic
Odor: agreeable
Fruiting body: (20-25 x15)cm, exoperidium white with reticulated mosaic, or smooth when young, (figures a,b) changing darker when mature, splitting open to release spore mass, without sterile base (figure c),
Spore mass: cheesy white when young, and powdery brown when mature (figure c)
Photographic sample only
GMA PT 2 PSA 5.18F
3915m   15°43'17.388" S 70°9'31.896" W
Observations: *Calvatia* (Arora 1986:634): conforms with descriptions of *Calvatia pachydermica* (Quispe Pelaez, 2020). Characterized by large carpophore size and reticulated peridium and brown spore mass. Locally traditionally consumed. Quechua 'Paku'

Cusco, Cuzco, Peru

© Peter Trutmann

*cf. **Calvatia sp.***
Northern Andes: Cajamarca

Spores: x̄ (3.7x3.7) µm (Q=1.0)

© Peter Trutmann

© Peter Trutmann

© Peter Trutmann

1 unit = 3.75 µm

1 unit = 15 µm

x 20

© Peter Trutmann

Jaén Province, Cajamarca, Perú

Region: Colasay, Jaen, Cajamarca - Northern Andes (figure i)
Ecology: organic material in endangered relic high selva northern rainforests (fig. h)
Macroscopic
Odor: none
Fruiting body: 4-5 cm high, 5 cm diameter, peridium white with small brown feltlike
spotted covering, with small dark patches with setae or hairlike growth, without
apical pore, with 1 cm tall sterile base (red arrow) (figures a,b,c)
Spore mass: mustard yellow
Microscopic
Spores: (3.4-4.1)x(3.4-4.1) µm, x̄ (3.7x3.7) µm (Q=1.0) (n=5), globose, lightly
ornamented with short spines, without pedicel, hyaline(KOH) (figure d)
Basidia: not detected
Capillitium: with sparsely branched hyaline hyphae (figures e,f :black arrows)
Cortical section: No chambers detected and without pedicels (figure g: white arrow)
Sample: GMA PT 5 PNA 3.5
est. 2435m   5°58'10.999" S 79°6'18" W
Observations: *cf. Calvatia* **(**Arora 1986:634**)**: it lacks peridioles. Lacks internal
spore mass chambers like *Pisolithus*. Possibly *Mycenastrum* due to feltlike surface.

P57　*cf. **Calvatia sp.***　　　　Spores: x̄ (6.3x6.2)μm (Qm=1.0)
Northern Andes: Tumbes

Region: Tumbes, Tumbes - Northern Andes (figure g)
Ecology: sandy soil below a native leguminous tree in northern semi
dry tropical grasslands in rainy season (figure g)
Macroscopic
Fruiting body: (5-7.5)cm high, (4x10) cm wide, exoperidium metallic
brown when mature, with small black marks, and minimal sterile
base (figures a,b)
Spore mass: cottony to powdery light brown (figures b,c)
Microscopic
Spores: (5.6-7.1)x(5.6-7.1)μm, x̄ (6.3x6.2)μm (Qm=1.0) (n=5),
globose, ornamented with short spines, without pedicel, green-
brown (KOH) (fig. d)
Capillitium: with pigmented, infrequently branching hyphae (figures
e,f: red arrows)
Sample　GMA PT 5 PNA 5.5
68 m　　　3°45'23.363" S 80°23'9.437" W
Observations: *cf. Calvatia* (Arora 1986:634): because of size and
description but lacks sterile base. It is probably not a large *Bovista*
species because spores lack pedicels.

1 unit = 3.75 μm

1 unit = 15 μm

1 unit = 3.75 μm

Disciseda

Disciseda candida   Spores: x̄ (4.0x4.0)μm c.(Qm=1.0)
Northern Andes: Cajamarca

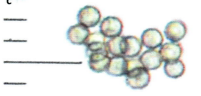

1 unit = 3.75 μm

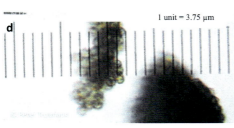

Region: Cerro el Castillo, Cajamarca (figure e)
Ecology: soil and moss under shrubs by degraded pastures (figure c, e)
Macroscopic
Fruiting body: (2x1)cm, spherical somewhat flattened, dark brown exoperidium rupturing to partly over (orange arrows) the underlying white endoperidium, with apical pore, no sterile base (figure b)
Spore mass: brown
Microscopic
Spores: (3.7-4.5)x (3.7-4.5)μm, x̄ (4.0x4.0)μm c.(Qm=1.0) (n=5), globose, lightly ornamented, greenish brown (KOH), without pedicel (figure c)
Capillitium: no mycelium found (figure d)
Sample: GMA PT 3 PNA 3.4
3192 m   7°13'47.166" S 78°29'14.915" W
Observations: *Disciseda*: *D. candida* (Schwein.) Lloyd (Dennis, 1970): p.10-11. which differs only by not having spores with short stubby pedicels (Arora 1989: 689). Found also in California and Argentina.

Lycoperdon

P59 **_Lycoperdon_ sp.**   Spores: (3.5-4)µm (Qm=1.0)
Southern Andes: Puno

5 µm

25µm

Region: Lampa, Puno- Southern Andes (figure e)
Ecology: soil in the highland puna grasslands (figure e)
Macroscopic
Fruiting body: 1-3cm, nearly round to pear-shaped,
exoperidium white warted with pyramidal spines without
apical pore (figures, a, b)
Sterile base: (0.2-0.5)cm spongy
Spore mass: brown
Microscopic
Spores: (3.5-4)µm, $\bar{x}$ (3.7)µm (Qm=1.0) (n=5), globose, thick
walled, ornamented lightly with spines, brown (KOH), with no,
or very short, pedicel (figures c, d)
Capitillium: with branched melanated hyphae (2µm) (figure d)
Sample: GMA PT 1 PSA 1.2
3823 m   15°18'47.55" S 70°12'41.844" W

Observations: _Lycoperdon_ (see Arora 1986:690 and Dennis
1970:10-11): Possibly L. _wrightii_ ?

Lycoperdon sp.

Spores: (3.4-3.8)μm (Qm=1.0)

Central and Northern Andes: Ancash, Cajamarca

1 unit = 3.75 μm

Region: Carhuaz, Ancash and Cajamarca, Cajamarca- Central and Northern Andes (figure f)

Ecology: soil in grass below shrubs in disturbed and degraded environment (figure a, e)

Macroscopic

Fruiting body: (1-1.5)x(2-2.5)cm, round to pear-shaped, peridium with white endoperidium and exoperidium forming an ornamented mosaic with superficial spinelike protrusions, with an apical pore (figures b,c)

Sterile base: 0.5-1cm, white, spongy, well developed (figure c)

Spore mass: light brown

Microscopic

Spores: (3.7)μm, x̄ (3.7)μm (Qm=1.0) (n=5), globose, smooth to very lightly ornamented, green(KOH), without pedicel (figure d)

Capillitium: not found (disintegrating with age?)

Sample:

Carhuaz, Ancash: GMA PT 3 PNA 8.7
2513 m   9°13'9.624" S 77°41'14.453" W
Cajamarca, Cajamarca: GMA PT 3 PNA 3.3
3190 m   7°13'47.4" S 78°29'14.73" W

Observations: *Lycoperdon* (Arora 1986:690): with finer and superficial spines than GMA PT 1 PSA 1.2

© Peter Trutmann

Lycoperdon sp.

Not yet evaluated microscopically

Southern Andes: Cusco

Region: Canchis, Cusco - Southern Andes (figure d)
Ecology: soil in grass in agro-pasture (figure d)
Macroscopic
Odor: N.E.
Fruiting body: (3-4)x(3-4)cm, white, with exoperidium
forming fine white fragile spinelike protrusions almost like a
powder (figure
Sterile base: 1 x1cm
Spore mass: brown
Microscopic
Pending
Sample: GMA PT 1 PSA 3.8
3406 m   13°58'19.776" S 71°29'36.503" W
Observation: *Lycoperdon* (see Arora 1986:690 and Dennis
1970:10-11). The specimen is similar to GMA PT 3 PNA 8.7
and GMA PT 3 PNA 3.3 from Ancash and Cajamarca, but
fruiting body specimens are larger.

cf. **Lycoperdon sp.**   Spores: (3.7x4.1)μm, (Qm=1.0)
Southern Andes: Arequipa

Region: Arequipa, Arequipa - Southern Andes (figure d)
Ecology: soil, highland puna grasslands under low shrubs (figure d)
Macroscopic
Odor: N.E.
Fruiting body: (3x2x1)cm, oval - cupcake like, peridium white, smooth with fine flannel like surface, with apical pore (figures a,b)
Sterile base: 1cm high x 2cm wide, white, cylindrical (figure b)
Spore mass: yellow-yellow brown
Microscopic
Spores: (3.7x4.1)μm, x̄ (3.9)μm (Qm=1.0) (n=5), globose, smooth or very lightly ornamented, without pedicel (figure c)
Capillitium: not found
Sample: GMA PT 3 PSA 8.8
3925 m   16°4'6.168" S 71°33'4.661" W
Observation: placed in *Lycoperdon* (Laessoe and Petersen, 2019): p 1229).

1 unit = 3.75 μm

P63   *Lycoperdon sp*   Spores: x̄ (3.6x3.6)μm (Qm=1.0)
       Northern Andes: Cajamarca

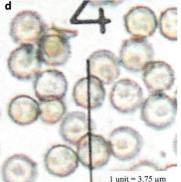

1 unit = 3.75 μm

Region: Cajamarca, Cajamarca (figure e)
Ecology: on dung on very poor soil of highland native puna grasslands
(figures a,e)
Macroscopic
Odor: N.E.
Fruiting body: (3-4) x (2-3)cm, exoperidium ornamented with a fine brown
mosaic over a white endoperidium, without apical pore (figure b,c)
Sterile base: reduced 0.5cm, white (figure c)
Spore mass: brown
Microscopic
Spores: (3.3-3.7)x(3.2-3.7)μm, x̄ (3.6x3.6)μm (Qm=1.0) (n=5), globose
green-brown (KOH), lightly ornamented with short spines, without pedicel
(figure d)
Capillitium: not found
Sample: GMA PT 3 PNA 4.5
3654 m   7°11'17.165" S 78°34'36.054" W
Observations: *Lycoperdon* (Dennis 1970:11) and Arora (Arora 1989:690).

cf. **Lycoperdon**
Southern Andes: Puno

Microscopic analysis pending

Region: Pucara y Lampa Puno - Southern Andes (figure d)
Ecology: soil, in highland agro-pastures (figure d)
Macroscopic
Fruiting body: 1-4cm. peridium white when young changing
to yellow-brown when mature, with superficial wart-like
granules, without apical pore (figures a,b,c),
Sterile base: (0.2-0.5)cm , small, (figure c)
Spore mass: yellow-brown
Microscopic
Pending
Sample:
Pucara, Puno: GMA PT 1 PSA 2.3
3918 m   15°2'40.074" S 70°22'20.459" W
Lampa, Puno: GMA PT 2 PSA 7.7 3844 m
15°18'46.337" S 70°12'42.767" W
Observations: *Lycoperdon*: possibly *L. marginatum* (see
Arora 1986:694 peeling puffball).

cf. **Lycoperdon sp.**
Southern Andes: Puno

Pending microscopic analysis

Region: Sillustani, Puno - Southern Andes (figure d)
Ecology: soil in highland degraded puna grasslands (figure d)
Macroscopic
Odor. N.E.:
Fruiting body: (1-3)x2cm, peridium smooth white with lumps
and pustules when young, peridium changing to brown
when mature with exoperidium splitting to produce a rough
mosaic pattern over a smooth brown endoperidium, with
apical pore (figure b)
Sterile base: reduced (0.2-0.5)cm (figure c)
Spore mass: brown
Microscopic
Pending
Sample: GMA PT 1 PSA 9.2
3875 m   15°43'21.372" S 70°9'21.089" W
Observation: *Lycoperdon* (Arora 1986: 690)

Lycoperdon *cf.* **perlatum**   Photographic record only
Southern Andes: Apurimac

Region: Abancay, Apurimac - Southern Andes (figure c)
Ecology: organic material and leaf litter in a highland *Pinus radiata* plantation (figure c)
Macroscopic
Odor: strong
Fruiting body: 1-3cm, spherical ball on cylindrical base, white when young, when mature peridium cream, exoperidium forming small brown superficial spines above cream endoperidium, with apical pore when mature (figures a,b)
Sterile base: 1-2cm high x 1 cm wide, highly developed into a cylindrical stem-like structure, cream and also with spines (figure b)
Spore mass: brown
Microscopic
Only photographic record from the field
Photographic sample: GMA PT 2 PSA 3.4F
2849 m   13°33'11.579" S 72°48'44.621" W
Observation: *Lycoperdon:* conforms to description of *Lycoperdon perlatum* (Arora 1989: 693). Common in pine woods of North America.

Podaxis

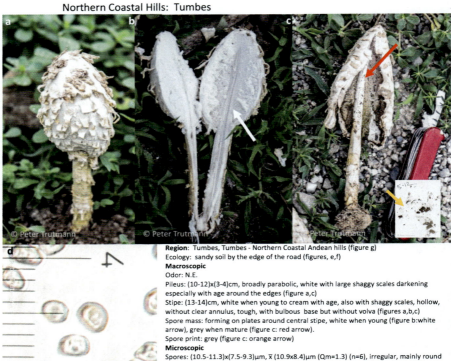

P67   ***Podaxis* *cf. pistillaris***   Spores: x̄ (10.9x8.4)μm (Qm=1.3)
Northern Coastal Hills: Tumbes

Region: Tumbes, Tumbes - Northern Coastal Andean hills (figure g)
Ecology: sandy soil by the edge of the road (figures, e,f)
Macroscopic
Odor: N.E.
Pileus: (10-12)x(3-4)cm, broadly parabolic, white with large shaggy scales darkening especially with age around the edges (figure a,c)
Stipe: (13-14)cm, white when young to cream with age, also with shaggy scales, hollow, without clear annulus, tough, with bulbous base but without volva (figures a,b,c)
Spore mass: forming on plates around central stipe, white when young (figure b:white arrow), grey when mature (figure c: red arrow).
Spore print: grey (figure c: orange arrow)
Microscopic
Spores: (10.5-11.3)x(7.5-9.3)μm, x̄ (10.9x8.4)μm (Qm=1.3) (n=6), irregular, mainly round to broadly elliptical or pear shaped, smooth, hyaline (KOH), sometimes clearly with germ-pore (figure d)
Capilililitium: N.E.
Pileipellis: N.E.
Sample: GMA PT 5 PNA 5.12
46 m   3°44'20.555" S 80°24'34.932" W
Observations: *Podaxis*: Dennis (1970:8-9,) Arora (1986:725). Similar to *P. pistillaris* who's lower end spore size is larger (10-16)x(9-15)μm, but specimen mean fits. Upper end spore size is too large for *P. microsporus* (5-9)μm.

1 unit = 3.75 μm

Tulostoma

P68F **Tulostoma** *cf.* **macrocephalum**   Photographic record only
Southern Andes: Puno

Region: Ilave, Puno - Southern Andes (figure d)
Ecology: soil on a bank beside the main road close to Pucará
(figure d)
Macroscopic
Fruiting body: (1-2)cm, spherical to oval head above tapered stem,
white with thin brown remnants of what appears to be an
exoperidium, with apical pore (figure a)
Stipe: (5-6)cm, white, tapered, underground (figure b)
Spore mass: brown (figure c)
Photographic sample
GMA PT 2 PSA 7.11F
3851 m   15°18'50.25" S 70°12'41.609" W
Observation: *Tulostoma*: see Dennis (Dennis 1970:2 and 14) and
Arora (Arora 1986:16-6 and 720).

Cystoderma

P69 **Cystoderma** *cf.* ***amianthinum***   Spores: x̄ (5.6x2.9)μm (Qm=2.8)
Central Andes: Cerro de Pasco

in Melzer's

1 unit = 1.5 μm

1 unit = 3.75 μm   x40

Region: Vicco, Cerro de Pasco (figure h)
Ecology: organic matter in Ichu, puna grasslands (figure g)
Macroscopic
Pileus: 2-4 cm, broadly conical to umbonate, yellow, smooth (figure a)
Stipe: 3-6cm, yellow, with scales ending in an irregular ring above
which the surface is smooth, without a volva (figures a,b)
Gills: white, emarginate, close
Spore print: white (figure b: orange arrow)
Microscopic
Spores: very small, (4.5-6.7)x(2.3-3.8) x̄ (5.6x2.9)μm (Q=1.9) (5), ,thin
walled, hyaline (KOH) slightly amyloid (or pseudoamyloid) with
Melzer's (red arrow), possibly slightly ornamented, no pore (figure c)
Trama: parallel (figure e)
Cistidia: none found
Basidia: pseudoamyloid, c. (25 x 3) μm (figures c,e)
Pileipellis: dermis: (globular epithelium?) of phaerocysts (white
arrow)with rusty brown walls (KOH)–radial section (figure d, f)
Sample: GMA PT 7 PCA 2.11
4152 m   10°50'26.91" S 76°10'37.308" W
Observations: Similar to *C. amianthinum* (Scop.) Fayod with spores (4-
7)x(3-4) (see Mushroom expert), but not to descriptions in Dennis
1970 p.58, now *C. jasonis* (5-6.5)x (3-3.5)μm (Venezuela). Family
status is in doubt. A new proposal places Cystodermateae in the
Squamanitaceae (D. Newman pers.com).

Peru

Chamaemyces

P70　　*cf.* **Chamaemyces sp.**　　　　　Spores: x̄ (7.4x5.0)μm (Qm=1.5)

Southern Andes: Arequipa

Region: Arequipa, Arequipa - Southern Andes (figure h)

Ecology: organic material below native bushes in dry sandy native puna grassland (figure g)

Macroscopic

Odor and taste: spicy-peppery

Color reaction: none

Pileus: 7cm, plane, white, fleshy, smooth, dry (figures a, b)

Stipe: 5(1)cm, white, stout, smooth, central, membranous, intermediate type annulus, without volva (figure c)

Gills: white, free, crowded

Spore print: did not yield a print (presumed white-off white)

Microscopic

Spores: (7-7.5)x(4.5-5.5)μm, x̄ (7.4x5.0)μm (Qm=1.5) (n=5), ellipsoid-subglobose, smooth or possibly punctate (finely spinose?), thin walled, hyaline (KOH), inamyloid (Melzer's), without germ-pore (figures, d,e)

Pileipellis: Unclear dermis or cutis of repent looking, inflated, brick-like cells within gelatinous medium - scalp section (figure f)

Sample: GMA PT 3 PSA 8.7

3922 m   16°4'6.168" S 71°33'4.637" W

Observations: Keys out to *Chamaemyces* (Singer 1975:469). Terricolous or decayed wood. Assuming smooth spores (if spores not smooth, but puntate or spinose then like *Smithomyces*, although no distinct volva was found).

1 unit = 3.75 μm

Chamaemyces sp.   Spores: x̄ (8.6x5.9)µm (Qm=1.4)
Southern Andes: Puno

a   b   c

© Peter Trutmann   © Peter Trutmann   © Peter Trutmann

d   e

f

1 unit = 3.75 µm

Region: Huancane-Putina, Puno - Southern Andes (figure h)
Ecology: soil, in ichu based puna highland grasslands (figure g)
Macroscopic
Odor: N.E.
Color reaction: none
Pileus: 4.5cm, broadly convex, smooth white finely speckled light brown, darkening to a cream or light brown at center, with eroded edges (figure a, b)
Stipe: 5(1)cm, white, stout, with membranous intermediate annulus, clavate, without volva (figure b)
Gills: with to light cream, free, crowded (figure c)
Spore print: white? (uncertain).
Microscopic
Spores: (7.5-11.2)x(5.6-6.7)µm, x̄ (8.6x5.9)µm (Qm=1.4) (n=6), elliptical, smooth, hyaline (KOH), inamyloid (Melzer's), with thick cell wall, and without germ-pore (figure d)
Trama: not evaluated
Cystidia: Pileicystidia (figure f)
Basidia: c.(32x6)µm, narrow and long, hyaline (KOH) with 2 to 4 sterigmata (fig. f: red arrow),
Pileipellis: a dermis with what looks like an epithelium of large digitate, many with pigmented cells (KOH) - pileocystidia c(65-70)µm growing from smaller palisade cell base (figure f:white arrow) similar to those reported for *Chamaemyces fracidus*
Sample: GMA PT 2 PSA 9.1
3882 m   15°4'46.386" S 69°48'56.826" W
Observations: *Chamaemyces* (singer 1975:469). With inamyloid, smooth spores, a hymeniform epicutis an epithelium and white to cream spore print. Local Aymara name 'Kala Kala'

g   h

© Peter Trutmann   Bolivia

Cystolepiota

cf. **Cystolepiota sp.**   Spores: x̄ (4.0x2.6)μm (Q=1.5)
Northern Andes: Cajamarca

© Peter Trutmann

in Melzer's

1 unit = 3.75 μm

1 unit = 6.5 μm   1 unit = 3.75 μm

Region: Pucará, Cajamarca - Northern Andes (figure h)
Ecology: soil, below native trees in secondary forest (figure g)
Macroscopic
Odor: none
Pileus: 2cm, umbonate, covered with a flannel-downy surface composed of shades of brown colored scale fibrils, with a solid brown calotte, the outer layer sometimes peels or breaks showing the white base (figure a)
Stipe: 3cm, slender, also covered with a downy layer of brown fibrils on a white base, specimen had no annulus (figure b)
Gills: white, free, close (figure b)
Spore print: did not yield a print (assumed white)
Microscopic
Spores: (3.8-4.5)x(2.0-3.0)μm, x̄ (4.0x2.6)μm (Q=1.5) (n=5), elliptical hyaline (KOH, not amyloid (Melzer's) smooth, without germ-pore (figure c)
Trama: regular parallel, not amyloid (figure d: white arrow)
Cystidia: not found
Basidia: c.(12-16)x(4)μm, with two or more spores, lightly amyloid (figure d:red arrow)
Pileipellis: a dermis: (figure e: black arrow) with specialized inflated cells c. (45μm x 4) μm that provide the downy surface texture (figure f)
Sample: GMA PT 5 PNA 2.8
1267 m   6°4'50" S 79°8'45" W
Observations: Tentatively placed in *Cystoepiota*, because Singer (1975:471) refers as a non-amyloid spored Lepioteae with similar epithelial cells and a non-amyloid trama. The suprapellis is similar to that of *Lepiota brunneoincarnata*, a poisonous species found in Europe. The sample is also similar to *Cystoderma* by its non amyloid, not dextrinoid, small spores, but it does not have free gills. Not found in available keys and national and regional literature.

Lepiota

Lepiota sp.   Spores: x̄ (16.2x5.0)μm (Qm=3.2)
Northern Andes: Cajamarca

Region: Colasay, Jaen, Cajamarca - Northern Andes (figure h)
Ecology: soil close to a mother tree in primary relic Amazonian cloud forest (figure h)
Macroscopic
Odor: mild
Pileus: 5cm, umbonate, squamulose made of a multitude of small fragile white recurved scales, browning close to center, with solid brown calotte together forming a highly ornamented surface (figures a,b)
Stipe: 9cm, narrow, scabrous with fine white scales, specimen without annulus, without volva (figures a,b)
Gills: white, free, close (figure c)
Spore print: did not yield print (assumed white)
Microscopic
Spores: (15-16.9)x(4.5-5.6)μm, x̄ (16.2x5.0)μm (Qm=3.2) (n=6), signoid to fusiform with flattened abaxial side, hyaline (KOH, pseudoamyloid (Meltzer's) smooth, without germ-pore (figure d)
Trama: of enlarged elongated cells (figure g:yellow arrow)
Cystidia: possibly piriform pleurocystidia (cells appear larger than normal basidia) c. (30x15)μm and inamyloid (figure f: red arrow)
Basidia: c.(22-26)x(7-8)μm, inamyloid, with what appear to be 2 spores (figure g: white arrow)
Pileipellis: a dermis: (trichoderm?) with non repent elongated inflated hyphae - scalp section (figure e: black arrow)
Sample: GMA PT 5 PNA 3.3
2435m   5°58'11.29" S 79°6'17.64" W
Observations: *Lepiota* (Singer 1975: 445, 469): due to its smooth pseudoamyloid spores without germ-pores, absence of epithelium and trichdermis like pileipellis

1 unit = 3.75 μm

in Melzer's

in Melzer's

THE MACROFUNGI OF ANDEAN PERU Part 1

cf. **Lepiota** *sp.*   Spores: x̄ (5.4x3.5) µm (Qm=1.5)
Central Andes: Lima

© Peter Trutmann

in Melzer's

1 unit = 3.75 µm

© Peter Trutmann

Regiion: Zarate, Huarochirí, Lima - Central Andes (figure f)
Ecology: organic litter under a *Polylepis incana* cloud forest (figure e)
Macroscopic
Odor: N.E.
Pileus: 4-5cm, umbonate- plane, corrugate, surface white with pink - with calotte at center solid pink, dry (figure a)
Stipe: 7-8cm, red brown, firm, smooth with fine hairs, widening at base of the pileus, without evidence of an annulus, or volva (figure b)
Gills: white, free with prominent collarium, close-crowded (figure b)
Spore print: did not yield a print
Microscopic
Spores: (4.8-6.2)x(3.0-4.2), x̄ (5.4x3.5) µm (Qm=1.5) (n=5), elliptical, hyaline (KOH), amyloid (Melzer's), smooth, without germ-pore (figure c)
Trama: N.E.
Cystidia: N.E.
Basidia: N.E.
Pileipellis: a cutis: parallel horizontal hyphae above parenchyma like cells (black arrow) in a gelatinous layer - radial section (figure d)
Sample: GMA PT 6 PCA 1.14
2936 m   11°55'54.797" S 76°28'52.74" W
Observation: *Lepiota* (Singer 1975:469): an amyloid, smooth spored *Lepiota* with a cutis as pileipellis.

Huarochiri, Gobierno Regional de Lima, Perú

Peru

cf. **Lepiota sp.**   Spores: x̄ (8.7x5.1)μm (Qm= 1.7)
Central Andes: Cerro de Pasco

© Peter Trutmann

in Melzer's

1 unit = 1.5 μm

1 unit = 3.75 μm

Region: Yanacancha, Cerro de Pasco - Central Andes (figure d)
Ecology: organic material in *Polylepis besseri* cloud forest (figure d)
Macroscopic
Odor: N.E.
Pileus: 4cm, when mature umbonate, viscid, brown colored with streaks of lighter and darker brown above white background and dark central calotte (figure a)
Stipe: 6cm, light colored turning brown when old, hollow, without annulus (although it may be the result of age), without volva (figure b)
Gills: white, free, close
Spore print: did not yield print
Microscopic
Spores: (7.5-9.8)x(4.5-5.3), x̄ (8.7x5.1) μm, (Qm= 1.7) (n=5), ellipsoid, smooth, hyaline (KOH), dextrose (Melzer's), without germ-pore (fig. c)
Trama: divergent
Cystidia: cheiloocystidia found, but no pleurocystidia
Basidia: with 4 spores
Pileipellis: a dermis: a hymeniform dermis or cellular matrix (black arrow) in a gelatinous medium, some slightly pigmented - scalp section (figure d)
Sample: GMA PT 7 PCA 3.3
3734 m   10°38'17.909" S 76°10'14.412" W
Observation: *Lepiota* (Singer 1975:469): a cellular pileipellis with smooth, pseudoamyloid spores without germ-pores.

*cf. **Lepiota sp.***
Central Andes: Ancash

Spores: x̄ (8.4x4.7)μm, (Qm=1.7)

© Peter Trutmann   © Peter Trutmann   © Peter Trutmann

in Melzer's

1 unit = 3.75 μm

in Melzer's

in

Region: Carhuaz, Ancash (figure i)
Ecology: organic litter below shrubs in disturbed environment (figure h)
Macroscopic
Odor: strong agreeable
Pileus: (2.5-3.0)cm, convex-umbonate, dry, brown splitting into long linear sections above underlying white cap surface (figs. a,b)
Stipe: 5.5cm, white, sturdy, smooth, without annulus, nor volva (figure b)
Gills: white, noted as free, crowded (figure c)
Spore print: none (presumed white)
Microscopic
Spores: (7.5-10.5)x(3.8-5.6)μm, x̄ (8.4x4.7) μm, (Q= 1.7) (n=5), elliptical to subfusiform, hyaline (KOH), pseudoamyloid (Melzer's), smooth without germ-pore (figure d)
Trama: psurodamyloid (figures e, f).
Cystidia: what appear to be long, ventricose, or obclavate, very thin walled pleurocystidia (figure f: black arrow)
Basidia: c.(40 x 7)μm, 2-4 sterigmata (figure e: white arrow)
Pileipelllis: a dermis:(epithelium?)- scalp section (figure g)
Sample: GMA PT 3 PNA 8.2
2995 m   9°14'42.455" S 77°40'40.548" W
Observations: *Lepiota* (Singer 1975:469): hymeniform like pileipellis with smooth, pseudoamyloid spores without germ-pores.

© Peter Trutmann

P77   *cf. **Lepiota sp.***   Spores: x̄ (8.0x5.4)μm (Qm=1.5)
Northern Andes: Tumbes

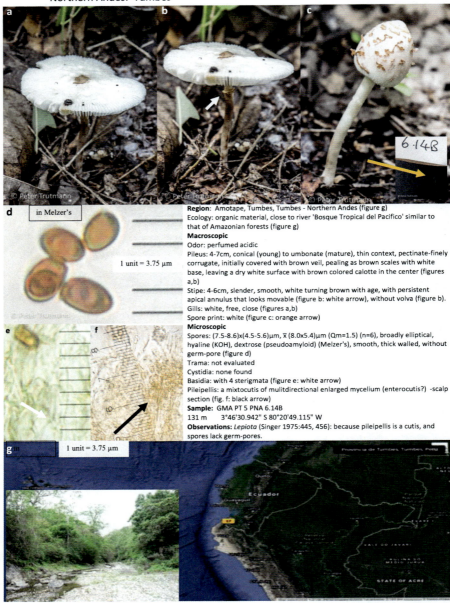

Region: Amotape, Tumbes, Tumbes - Northern Andes (figure g)
Ecology: organic material, close to river 'Bosque Tropical del Pacifico' similar to that of Amazonian forests (figure g)
Macroscopic
Odor: perfumed acidic
Pileus: 4-7cm, conical (young) to umbonate (mature), thin context, pectinate-finely corrugate, initially covered with brown veil, pealing as brown scales with white base, leaving a dry white surface with brown colored calotte in the center (figures a,b)
Stipe: 4-6cm, slender, smooth, white turning brown with age, with persistent apical annulus that looks movable (figure b: white arrow), without volva (figure b).
Gills: white, free, close (figures a,b)
Spore print: white (figure c: orange arrow)
Microscopic
Spores: (7.5-8.6)x(4.5-5.6)μm, x̄ (8.0x5.4)μm (Qm=1.5) (n=6), broadly elliptical, hyaline (KOH), dextrose (pseudoamyloid) (Melzer's), smooth, thick walled, without germ-pore (figure d)
Trama: not evaluated
Cystidia: none found
Basidia: with 4 sterigmata (figure e: white arrow)
Pileipellis: a mixtocutis of mulitdirectional enlarged mycelium (enterocutis?) -scalp section (fig. f: black arrow)
Sample: GMA PT 5 PNA 6.14B
131 m   3°46'30.942" S 80°20'49.115" W
Observations: *Lepiota* (Singer 1975:445, 456): because pileipellis is a cutis, and spores lack germ-pores.

in Melzer's

1 unit = 3.75 μm

1 unit = 3.75 μm

P78   *cf. **Lepiota sp.***   Spores: x̄ (6.3x3.5)μm (Qm=1.8)
Northern Andes: Cajamarca and Lambayeque

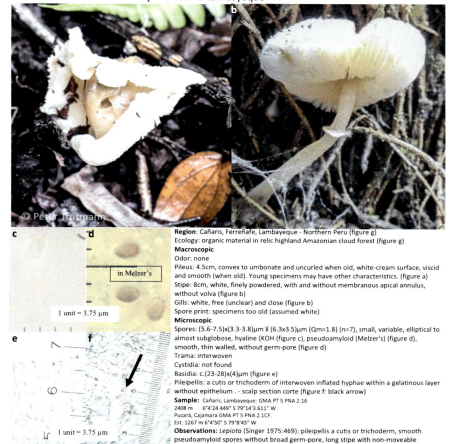

Region: Cañaris, Ferreñafe, Lambayeque - Northern Peru (figure g)
Ecology: organic material in relic highland Amazonian cloud forest (figure g)
Macroscopic
Odor: none
Pileus: 4.5cm, convex to umbonate and uncurled when old, white-cream surface, viscid and smooth (when old). Young specimens may have other characteristics. (figure a)
Stipe: 8cm, white, finely powdered, with and without membranous apical annulus, without volva (figure b)
Gills: white, free (unclear) and close (figure b)
Spore print: specimens too old (assumed white)
Microscopic
Spores: (5.6-7.5)x(3.3-3.8)μm x̄ (6.3x3.5)μm (Qm=1.8) (n=7), small, variable, elliptical to almost subglobose, hyaline (KOH (figure c), pseudoamyloid (Melzer's) (figure d), smooth, thin walled, without germ-pore (figure d)
Trama: interwoven
Cystidia: not found
Basidia: c.(23-28)x(4)μm (figure e)
Pileipellis: a cutis or trichoderm of interwoven inflated hyphae within a gelatinous layer without epithelium . - scalp section corte (figure f: black arrow)
Sample: Cañaris, Lambayeque: GMA PT 5 PNA 2.16
2408 m   6°4'24.449" S 79°14'3.611" W
Pucará, Cajamara GMA PT 5 PNA 2.1CF.
Est. 1267 m 6°4'50" S 79°8'45" W
Observations: *Lepiota* (Singer 1975:469): pileipellis a cutis or trichoderm, smooth pseudoamyloid spores without broad germ-pore, long stipe with non-moveable annulus, Specimens in poor condition, old and consumed by maggots.

in Melzer's

1 unit = 3.75 μm

1 unit = 3.75 μm

cf. ***Lepiota sp.***   Spores: x̄ (7.5x3.8)µm (Qm=2.0)
Northern Andes: Lambayeque

© Peter Trutmann   © Peter Trutmann   © Peter Trutmann

in Melzer's

1 unit = 1.5 µm

1 unit = 3.75 µm

Region: Ferreñafe, Lambayeque - Northern Andes (figure f)
Ecology: organic material under shrubs in secondary growth (figure f)
Macroscopic
Pileus: 1-2cm, umbonate, corrugate, sometimes splitting with age, thin fleshed, white with orange scales calotte at center darker and solid (figure a, b)
Stipe: 2-3cm, white with sparse large sales pealing from surface, with fragile annulus, without volva (figure b)
Gills: white, free, close, close (figure c)
Spore print: did not yield a print
Microscopic
Spores: (6.8-8.3)x(4.2-5.3)µm, x̄ (7.5x3.8)µm (Qm=2.0) (n=5) elliptical, smooth, hyaline(KOH), pseudoamyloid (Melzer's), without germ-pore (figure d)
Trama: not determined, but reacting to dextrose to Melzer's
Cystidia: not found
Basidia: not found
Pileipellis: a cutis: of parallel pigmented hyphae perhaps above a cellular base (figure e: black arrow) covered with melanated setae without septa c. (150µm o mas x 7.5µm) - perhaps of an epiphytic colonizer - scalp section (figure e: red arrow)
Sample: GMA PT 6 PNA 2.7
1449 m   6°1'54.647" S 79°12'8.339" W
Observations *Lepiota* (Singer 1975:469): a mycelial pileipellis, with smooth, pseudoamyloid spores without germ-pores. Not found in Dennis (Dennis, 1970) or Arora (1986), or nationally (Alvarez Loayza et al., 2014; Laessoe T. and Petersen J.H., 2008; Mata et al., 2006) Cárdenas et al (Cárdenas Medina et al., 2019). Different from P80 nearby in Cajamarca by presence of an annulus, texture of the cap and stipe and spores (Qm =2 vs 1.5). Not *Leucoagaricus rubrotinotus as it has* a different pileipellis.

Lepiota s.l.   Spores: x̄ (7.1x4.3)μm (Qm=1.5)
Northern Andes: Cajamarca

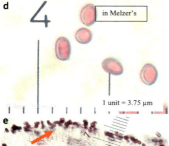

in Melzer's

1 unit = 3.75 μm

in

Region: Pucará, Cajamarca - Northern Andes (figure g)
Ecology: organic debris beneath secondary native forest (figure f)
Macroscopic
Odor: chemical-aromatic
Pileus: 1.5cm, umbonate, corrugated, orange-red covering when very young, splitting
to a long red-orange streaks over white base , with center calotte remaining darker
(figure a)
Stipe: 3cm, white, smooth, without annulus of volva (figures, b,c)
Gills: white, free, close (figure c)
Spore print: not taken (only one specimen found)
Microscopic
Spores: (6.9-7.5)x(3.8-4.5)μm, x̄ (7.1x4.3)μm (Qm=1.5) (n=5) elliptical, hyaline (KOH) ,
pseudoamyloid (Melzer's) smooth, thick walled, without germ-pore (figure d)
Trama: enlarged pseudoparaphyses (pavement) cells, non-amyloid in Melzers (figure
e: black arrow)
Cystidia: none observed
Basidia: c.(11x4)μm, not amyloid, with 2 or more spores (figure e: red arrow)
Pileipellis: uncertain
Sample: GMA PT 5 PNA 2.9
1256 m   6°4'50.022" S 79°8'45.072" W
Observations: *Lepiota s.l* (Singer 1975:469): due to small size, thin cap,
psuedoamiloid spores without germ-pores and stipe without annulus. Cresyl blue
metachromatic stain of spores not performed. If metachromatic then
Leucagaricus. Visually similar to *Leucoagaricus rubrotinctus* (Peck) Singer 1948, (see
P94 a,b) as is spore size (6-9)x(4-5)μm, but carpophore is smaller, lacks an annulus,
lacks cheilocystidia and spore form differs.

P81 **_Lepiota_** _s.l._   Spores: x̄ (8.8x5.1)μm (Qm=1.6)
(2)   Southern Andes: Apurimac

© Peter Trutmann

© Peter Trutmann

in Melzer's

1 unit = 3.75 μm

in Melzer's   1 unit = 6.5 μm

Region: Abancay, Apurimac (figure f)
Ecology: organic material in Podocarpus glomeratus forest (figure e)
Macroscopic
Color reaction: none (slight brown)
Pileus: 2-3cm, umbonate, cream to light pink, calotte at center, margin slightly
uplifted and undulated, slightly moist, smooth, flesh white (figure a)
Stipe: 4-7cm, white, bruising slightly brown, smooth, with delicate membranous veil,
slightly bulbous at foot but without volva (figures a,b)
Gills: white, free, crowded (figure b)
Spore print: did not yield print (presumed white)
Microscopic
Spores: (7.0-9.5)x(4-5.5)μm, x̄ (8.2x5.1)μm (Qm=1.6) (n=10), elliptical, hyaline (KOH),
dextrinoid (Melzer's: white arrow), smooth, no germ-pore (figure c)
Trama: interwoven - no large peudoparaphyses (pavement cells) characteristic of
Leucocoprinus (figure d: black arrow)
Cystidia: not found
Basidia: with 2-4 spores (figure d)
Pileipellis: not evaluated
Sample: GMA PT 3 PSA 2.5
3029 m   13°36'0.779" S 72°52'35.249" W
       GMA PT 3 PSA 2.13
3191 m   13°35'52.877" S 72°52'42.185" W
Observation: Tentatively Lepiota s.l. (Singer 1975:469): due to form, smooth,
pseudoamyoid spores without germ-pores, and non persistent non movable
annulus. Pileipellis information is required.

P82   *cf. **Lepiota sp.***   Spores: x̄ (7.2x5.0)μm, (Qm=1.4)

Central Andes: Ancash

Region: Carhuaz, Ancash - Central Andes (figure l)
Ecology: organic material below shrubs (figure k)
Macroscopic
Odor: mild
Pileus: 1-2cm, conical, umbonate to mammilate, white with light brown central calotte, smooth, dry (figures a,b,e)
Stipe: 3-4 cm, white, smooth, with central to apical membranous annulus, without volva (figures, b, c)
Gills: white, free, crowded (figure f)
Spore print: did not yield print (presumed white)
Microscopic
Spores: (6.8-9.5)x(3.8-5.6)μm, x̄ (7.2x5.0)μm, (Qm=1.4) (n=5), ellipsoid, hyaline (KOH), dextrose (Melzer's), smooth, without germ-pore (figure g).
Trama: N.E.
Cystidia: none observed
Basidia: with 4 or more spores (figure h: white arrow)
Pileipellis: a dermis -hymeniform or complex cellular matrix (figure j) with cells that produce pigmented, c. 4μm long crystals (figures i,j: black arrow)
Sample: GMA PT 3 PNA 8.1
2995 m 9°14'42.455" S 77°40'40.548" W
Observation: *Lepiota* (Singer 1975:445,469): by carpophore form, spores that lack germ-pores, lack of persistent and movable annulus and a cellular dermis like form as epicutis. A cresyl blue metachromatic stain of spores was not performed. If metachromatic then *Leucocoprinus*.

in Melzer's

1 unit = 1.5 μm

1 unit = 7.5 μm

1 unit = 3.75 μm

Lepiota s.l.   Spores: x̄ (8.4x5.7)μm (Qm=1.5)
Southern Andes: Cusco

Region: Zurite, Cusco- Southern Andes (figure e)
Ecology: organic matter, in *Polylepis racemosas* and *Polylepis besseri* cloud forest (figure e)
Macroscopic
Pileus: 3cm, umbonate to mammilate, brown to pinkish brown, with warted scales and white base darker at the center calotte (figures a,b)
Stipe: 6cm, white changing to cream and dark brown, with apical fragile membranous annulus, without volva (figures, b,c)
Gills: white, free, close (figures b,c)
Spore print: did not yield print (assumed white)
Microscopic
Spores: (7.9-9.5)x (4.5-6.5)μm, x̄ (8.4x5.7)μm (Qm=1.5) (n=5) elliptical, smooth, thin walled, hyaline (KOH), dextrose (Melzer's), without germ-pore (figure d)
Sample: GMA PT 3 PSA 7.7
3880m   13°25'58" S 72°15'17" W
Observations: Tentatively *Lepiota* s.l. (Singer 1975:469): by form, smooth, pseudoamyoid spores without germ-pores, as well as non persistent non movable annulus. Pileipellis information lacking.

d   1 unit = 3.75 μm   in Melzer's

Lepiota s.l.

Spores: x̄ (7.5x5.2)μm (Qm=1.4)

Central Andes: Ancash

in Melzer's

1 unit = 3.75 μm

Region: Carhuaz, Ancash - Central Andes (figure e)
Ecology: organic litter under shrubs in disturbed environment (figure d)
Macroscopic
Odor: strong but agreeable
Pileus: 2-3cm, broadly convex, orange, viscid, smooth, with overhanging undulating edges (figure a)
Stipe: 7-7.5cm, white, smooth, hollow, with intermediate type annulus, without volva (figure a)
Gills: white, free, crowded (figure b)
Spore print: no yield (assume white)
Microscopic
Spores: (7.0-8.3)x(4.8-5.6)μm, x̄ (7.5x5.2)μm,(Qm=1.4) (n=5), elliptical, hyaline (KOH), dextrose (Melzer's), smooth, without germ-pore (figure c)
Trama: N.E.
Cystidia: N.E.
Basidia: N.E.
Pileipellis: N.E.
Sample: GMA PT 3 PNA 8.4
2996 m   9°14'44.61" S 77°40'29.837" W

Observations: Tentatively *Lepiota* s.l. (Singer 1975:469) - rather than *Leucoagaricus*: due to its fragile stature, and smooth, pseudoamyloid spores without germ-pores, and non movable annulus. Pileipellis information lacking.

Lepiota s.l.   Spores: x̄ (14.4x5.9)µm (Qm=2.4)
Southern Andes: Apurimac

a
b
© Peter Trutmann
© Peter Trutmann

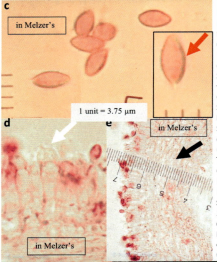

c
in Melzer's
1 unit = 3.75 µm
d
in Melzer's
e
in Melzer's
in Melzer's

Regiion: Abancay, Apurimac (figure f)
Ecology: organic material under *Polycarpus glomeratus* (figure f)
Macroscopic
Odor: strong like Porcini mushrooms
Pileus: 1-2.5 cm, convex to umbonate, white with brownish red scales darker in central calotte (figure a)
Stipe: 3-5 cm white, with filamentous white stripes, and apical superior white to cream membranous annulus, without volva (fig. b)
Gills: white, free, crowded (figure b)
Spore print: white
Microscopic
Spores: (11.5-15.0)x(5.5-7.0)µm, x̄ (14.4x5.9)µm (Qm=3.4) (n=5), elliptical to subfusiform, lightly ornamented (rugose) (red arrow), hyaline (KOH), red-pseudoamyloid (Melzer's), without germ-pore (fig. c)
Trama: made of inflated cells, parallel (figure e)
Cystidia: not found
Basidia: c.(40x10)µm with 4 sterigmata (figure c: white arrow)
Pileipellis: not evaluated
Sample: GMA PT 3 PSA 2.20
3240 m (est.)   13°35'50.334" S 72°52'42.258" W (est.)
Observations: *Lepiota* s.l. (Singer 1975:444,469): due to size and spores lacking prominent germ-pores. However, a cresyl blue metachromatic stain of spores was not performed. If metachromatic then a *Leucagaricus*. Like *P86* (GMA PT 3 PSA 2.25), but carpophore smaller, and spore shape different.

© Peter Trutmann

Lepiota s.l.   Spores: (9.7x5.2)μm (Qm=1.7)

Southern Andes: Apurimac

in Melzer's

1 unit = 3.75 μm

Region: Abancay, Apurimac (figure f)

Ecology: organic material under *Podocarpus glomeratus* (fig. e)

Macroscopic

Odor: agreeable

Pileus: 3.5-4cm, convex to broadly convex, first covered with brownish pink veil, then splitting to show white surface with brownish pink scales, center remaining solid color (figure a)

Stipe: 5-7cm, white, rigulose, changing to darker with age, with membranous apical annulus, tapered from base to apex, without volva (figures a, b)

Gills: white, free

Spore print: white

Microscopic

Spores: (9-11) x (5-7) μm, x̄ (9.7x5.2)μm (Qm=1.7) (n=5), elliptical almost subglobose, smooth, hyaline (KOH), pseudoamyloid (Melzer's), unclear if it has germ-pore (figure c)

Cystidia: not observed

Basidia: c.(24-28)x(8)μm), with numerous buds like protrusions that don't appear to be sterigmata (figure d: white arrow)

Pileipellis: not checked

Sample: GMA PT 3 PSA 2.25

3250 m (est.)   13°35'50.334" S 72°52'42.258" W (est.)

Observations: Tentatively *Lepiota* s.l. (Singer 1975:444,469): with smooth, pseudoamyloid spores that do not have germ-pores and a non-movable annulus. However, a cresyl blue metachromatic stain of spores was not performed. If metachromatic then it is a *Leucagaricus*. Pileipellis information is lacking.

Pseudobaeospora

P87   *cf.* **Pseudobaeospora sp.**   Spores: x̄ (8.3x4.5)μm (Qm=1.8)
Central Andes: Ancash

© Peter Trutmann   © Peter Trutmann   © Peter Trutmann

c

in Melzer's

1 unit = 3.75 µm

d

Region: Yungay, Ancash - Central Andes (figure f)
Ecology: organic matter in moss, in *Polycarpus glomeratus* cloud forest (fig. e)
Macroscopic
Odor: NA
Color reaction: none
Pileus: 1-15cm, umbonate, orange with darker center (calotte), smooth, dry (figures a,b)
Stipe: 4cm, white, fibrous, hollow, with fragile membranous annulus, bulbous but without volva (figures, a-d)
Gills: white, free, close (figure c)
Spore print: did not yield spores (assumed white)
Microscopic
Sproes: (7.5-9.4)x(4.5-4.9)μm, x̄ (8.3x4.5)μm (Qm=1.8) (n=5), elliptical, smooth, hyaline (KOH) pseudoamyloid (Melzer's), without germ-pore (figure c)
Trama: N.E.
Cystidia: N.E.
Basidia: N.E.
Pileipellis: Cutis a carpet of undifferentiated hyphae (figure d: black arrow) using 'scalp' section
Sample: GMA PT 5 PNA 1.4
3700m   9°5'25.476" S 77°40'41.07" W
Observation: Placed in the *Lepioteae* as *Pseudobaeospora* (Singer 1975:469,476): pileipellis a cutis, pseudoamyloid, smooth spores without germ-pore, and growing on ground, on humus and moss thali.

e

f

Chlorophyllum

P88   ***Chlorophyllum sp.***   Spores: (8.2x5.4)μm (Qm=1.5)
Southern Andes, Puno

Region: Huancho-Lima, Huancane, Puno (figure i)
Ecology: soil under shrubs, beside traditional farmhouse (figure g)
Macroscopic
Odor and Taste: not tested
Color reaction: no clear reaction
Pileus: 6-9 cm. lepinotoid, plane to broadly convex, white to cream, smooth, dry (figures a,b)
Stipe: 3-4 (0.5-1)cm, central, white, tough, with thick, intermediate type membranous sometimes moveable annulus, slightly bolbous base without volva (figures b,c)
Gills: white to lightly pink when older, free, crowded (figure c)
Spore print: colonial buff - light pink (figure c: orange arrow)
Microscopic
Spores: (6.5-10.0)x(4.5-6.5)μm, mean (8.2x5.4)μm (Qm=1.5) (n=6), elliptical, smooth, thick walled, green to light pink (KOH), not tested for Melzer', some with narrow germ-pore (fig.d: white arrow)
Trama: parallel (figures d,e)
Cystidia: not observed
Basidia: with 4 sterigmata (red arrow), and spores that appear coved by a membrane (figure f)
Pileipellis: Trichderm of a palisade of vertical greenish hyphae greenish with clamp connections (figure g: black arrow)
Sample: GMA PT 1 PSA 10.1
3857 m   15°10'33.306" S 69°50'58.608" W
Observations: *Chorophyllum:* by color and shape of spores and hyphae, structure of pileipellis (Singer 1975:454).

cf. *Chlorophyllum sp.*
Northern Coast: La Liberdad

Spores: x̄ (9.9x7.5)μm (Qm=1.3)

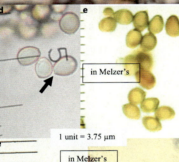

Region: Chan Chan, Trujillo, La Liberdad - Northern Coast (figure h)
Ecology: soil below native a stand of Huarango known also as 'algarrobo pálido' (*Prosopis pallida*) neara the entrance of Chan Chan (figure g)
Macroscopic
Odor: agreeable
Color reaction: redish brown (figure b: red arrow)
Pileus: 5-14cm, campanulate (young) to broadly conical (mature), dry, when young covered with remnants of dark brown veil above large white scales, white-cream when mature (figures a, c)
Stipe: 5-9(1-2)cm when young and 15cm when mature, fleshy, robust, white -cream, smooth with a movable one layered annulus, with bulbous base but without volva (figure b, d)
Gills: white to cream and brown when old, free (figures b,d,e)
Spore print: cream?
Microscopic
Spores: (8.2-12)x(6.7-8.2)μm, x̄ (9.9x7.5)μm (Qm=1.3) (n=11) Ovate to subglobose, smooth, thik walled, hyaline to greenish (KOH) (figure d), pseudoamyloid (Melzer's) (figure e), with narrow germ-pore without plug found in some spores (figures d: black arrow)
Cystidia: not found
Basidia: c.(28-30)x(7-8) μm, with 2-4 sterigmata (figures e,f)
Pileipellis: Not determined
Samples: GMA PT 2 PCN 2.4
GMA PT 2 PNA 2.6
30 m   8°6'1" S 79°4'1.999" W
Observations: cf. *Chlorophyllum* (Singer 1975: p.445): due to its carpophore shape, reddish discoloration, pseudoamyloid, green thick-walled spores that are less than <10μm with narrow germ-pores without plugs, a movable ring, absence of volva. It shares properties with *C. molybdites* usually with grey-olive-green spore print. It also shares characteristics with the edible *Leucoagaricus americanus*, in spore size (9-11 x 6-7 μm) and spore print color, but it does not stain reddish.

in Melzer's

1 unit = 3.75 μm

in Melzer's

Chlorophyllum esculentum
Northern Andes: Tumbes

Spores: x̄ (9.1x7.3)μm (Qm=1.2)

Region: Tumbes, Tumbes - Northern Andes (figure i)
Ecology: sandy soil in tropical grassland (figure h)
Macroscopic
Odor: agreeable
Color reaction: none
Pileus: (4-15)cm, campanulate when young to broadly conical- umbonate-plane, white to cream, fleshy, with remnants of the veil forming brown scales with a dark center and dry surface (figures a,b)
Stipe: 8-14(1)cm, thicker at base, white to cream, hollow, smooth, with membranous intermediate with double ring on stalk, becoming brownish on underside, without a volva (figure c)
Gills: white to greenish, free, crowded (figure c)
Spore print: greenish white (figure c: orange arrow)
Microscopic
Spores: (8.2-10.5)x(6.0-8.3)μm, x̄ (9.1x7.3)μm (Qm=1.2) (n=12) elliptical-subglobose, greenish (KOH, smooth, only sometimes with germ-pore (figure d), dextrose with Melzer's (figure f,g)
Trama: enlarged parallel cells (figure g: white arrow)
Cystidia: none observed
Basidia: c.(23-28)x(7-8) μm, with two or more sterigmata (figure f:red arrow)
Pileipellis: cutis or trichoderm: mycelium with some very elongated perpendicular structures (scalp section figure e: black arrow)
Samples: GMA PT 5 PNA 5.1
96 m   3°45'55.091" S 80°21'40.692" W
            GMA PT 5 PNA 6.1B
92 m   3°45'55.091" S 80°21'40.692" W
            GMA PT 5 PNA 5.10
51 m   3°44'29.988" S 80°24'33.005" W
Observations: *Chlorophyllum* (Singer 1975: 445): identified as *C. esculentum* Massee (7.5-10)x(5.5-7)μm. *C.molybdite* has been reported from similar semi-arid zone in North East Brazil (Neves et al., 2013) p.32 but it has larger (10-13)x(7-9)μm (Dennis, 1970:53), Arora (1986:295-7. *Chlorophyllum* has been reported as edible in South America (Singer 1975: p.455)

1 unit = 3.75 μm

in Melzer's   in Melzer's

Leucoagaricus

P91 **_Leucoagaricus sp._**   Spores: (6.9x5.2)µm (Qm=1.3)
(2)   Southern Andes: Cusco, Puno

in Melzer's

1 unit = 3.75 µm

Lactophenol
Cotton Blue

in Melzer's

Region: Huancane, Puno; Chinchero, Cusco - Southern Andes (figure J)
Ecology: soil in highland agro-pastures (figure k)
Macroscopic
Odor: agreeable
Color reaction: none
Pileus: 6-10cm, broadly convex, light brown, smooth, fleshy (figures,a,b,d)
Stipe: 3-10(1-3)cm, white, smooth, stout, central, with membranous
intermediate non movable annulus, without volva (figure a,b,c,e)
Gills, white, free, crowded (figure e)
Spore print: white-off white
Microscopic
Spores: (4.0-7.5)x(3.0-6.0)µm, x̄ (6.0x4.6)µm (Qm=1.3) (n=9), subglobose,
smooth, hyaline(KOH), pseudoamyloid (Melzer's), without germ-pore
(figures f,g)
Cystidia: not observed
Basidia: c. (30-40)x(7-8)µm (figure h), with 4 spores (black arrow)
pseudoamyoid (Melzer's) (figure i)
clamp connections not found
Pileipellis: a dermis:(trichoderm?): unable to determine from old specimen
(figure f)
Sample:
Huancane, PunoGMA PT 1 PSA 12.1
3839m   c. 15°10'47.254" S 69°50'37.846" W
Chinchero, Cusco GMA PT 3 PSA 3A.9
c. 3700 m   . 13°24'25" S 72°3'5.999" W
Observations: _Leucoagaricus_ (Singer 1975:445): has _Agaricus_ like
characteristics, smooth, pseudoamyloid spores (although germ-pore not
visible) and trichoderm like epidermis. Puno sample was old. Known in
Chinchero, Cusco as 'Papa Kallampa' (cultural, micrograph figure h, and
specimen in figure b provided by Milton Callañaupa UNSAAC, Cusco.(N.B
Much larger specimens have been found in Cusco)

Leucoagaricus cf. *nympharum*   Spores: (9.1x7.2) µm (Q=1.3)
Northern Andes: Tumbes

© Peter Trutmann   © Peter Trutmann   © Peter Trutmann

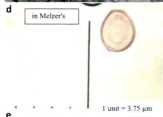

in Melzer's

1 unit = 3.75 µm

Region: Amotape, Tumbes, Tumbes - Northern Andes (figure g)
Ecology: soil on the side of the track below native trees (figure f)
Macroscopic
Odor: strong, but not disagreeable
Color reaction: no red reaction
Pileus: 5cm, plane, highly ornamented, with large upright white scales browning at the tips appearing like untidy hair (figure s a,b)
Stipe: 7 (0.5-1)cm, white, also covered with large white scales up to the annulus, the membranous non movable annulus also with scales sleeve-like hanging like peeled skin, above annulus stipe smooth and brownish (figures a,b,c)
Gills: white, free, crowded (figure c)
Spore print: did not yield a print (assumed white or off-white)
Microscopic
Spores: (8.2-10.1)x(7.1-7.5) µm x̄ (9.1x7.2) µm (Q=1.3) (6), truncated, broadly elliptical to subglobose, smooth, thick-walled, hyaline (KOH, pseudoamyloid (Melzer's), possibly with narrow germ-pore (unclear) (fig.d)
Cystidia: N.E.
Basidia: N.E.
Pileipellis: appears an epithelium with rounded cells dermis (a hymenidermis?) with palisade or possibly sphaerocyst-like cells (black arrow) in a gelatinous medium - scalp section (figure e)
Sample: GMA PT 5 PNA 5.9
70 m   3°45'25.566" S 80°23'28.248" W
Observations: *Leucoagaricus* (Singer 1975:445): by *Agaricus* like form of carpophore, with smooth, pseudoamyloid, truncated spores with germ-pore and dermis (hymenidermis?) like epicutis or possibly sphaerocysts. Fits description of L. nympharum (Laessoe and Petersen, 2019:340). Sample destroyed by mites before drying.

cf. Leucoagaricus sp.
Northern Andes: Cajamarca

Spores: x̄ (7.0x4.3)μm (Qm=1.6)

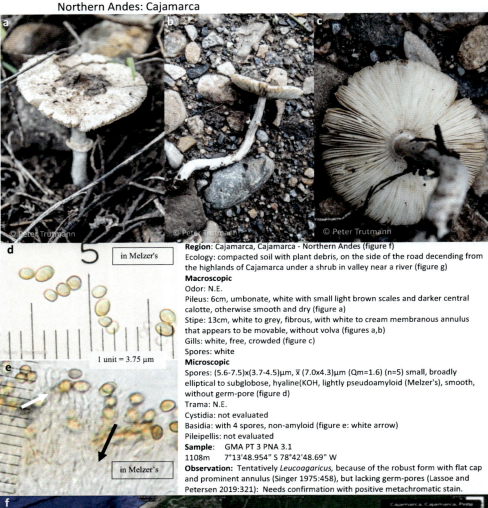

Region: Cajamarca, Cajamarca - Northern Andes (figure f)

Ecology: compacted soil with plant debris, on the side of the road decending from the highlands of Cajamarca under a shrub in valley near a river (figure g)

Macroscopic

Odor: N.E.

Pileus: 6cm, umbonate, white with small light brown scales and darker central calotte, otherwise smooth and dry (figure a)

Stipe: 13cm, white to grey, fibrous, with white to cream membranous annulus that appears to be movable, without volva (figures a,b)

Gills: white, free, crowded (figure c)

Spores: white

Microscopic

Spores: (5.6-7.5)x(3.7-4.5)μm, x̄ (7.0x4.3)μm (Qm=1.6) (n=5) small, broadly elliptical to subglobose, hyaline(KOH, lightly pseudoamyloid (Melzer's), smooth, without germ-pore (figure d)

Trama: N.E.

Cystidia: not evaluated

Basidia: with 4 spores, non-amyloid (figure e: white arrow)

Pileipellis: not evaluated

Sample: GMA PT 3 PNA 3.1

1108m   7°13'48.954" S 78°42'48.69" W

Observation: Tentatively *Leucoagaricus,* because of the robust form with flat cap and prominent annulus (Singer 1975:458), but lacking germ-pores (Lassoe and Petersen 2019:321): Needs confirmation with positive metachromatic stain.

(in Melzer's) — 1 unit = 3.75 μm — in Melzer's

P94a *Leucoagaricus cf. **rubrotinotus*** Spores: x̄ (7.5x4.3)μm (Qm=1.7)
Northern Andes: Tumbes

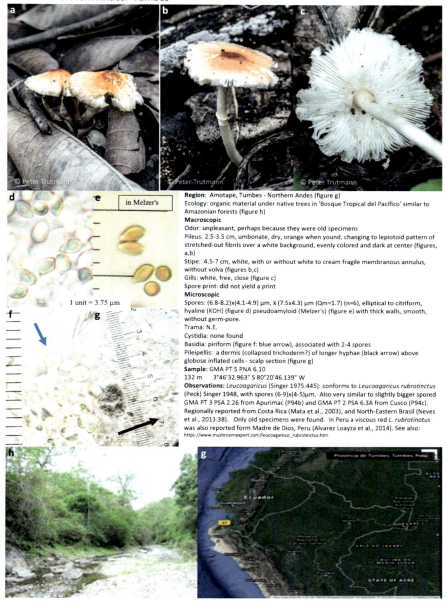

Region: Amotape, Tumbes - Northern Andes (figure g)
Ecology: organic material under native trees in 'Bosque Tropical del Pacifico' similar to Amazonian forests (figure h)
Macroscopic
Odor: unpleasant, perhaps because they were old specimens
Pileus: 2.5-3.5 cm, umbonate, dry, orange when yound, changing to lepiotoid pattern of stretched-out fibrils over a white background, evenly colored and dark at center (figures, a,b)
Stipe: 4.5-7 cm, white, with or without white to cream fragile membranous annulus, without volva (figures b,c)
Gills: white, free, close (figure c)
Spore print: did not yield a print
Microscopic
Spores: (6.8-8.2)x(4.1-4.9) μm, x̄ (7.5x4.3) μm (Qm=1.7) (n=6), elliptical to citriform, hyaline (KOH) (figure d) pseudoamyloid (Melzer's) (figure e) with thick walls, smooth, without germ-pore.
Trama: N.E.
Cystidia: none found
Basidia: piriform (figure f: blue arrow), associated with 2-4 spores
Pileipellis: a dermis (collapsed trichoderm?) of longer hyphae (black arrow) above globose inflated cells - scalp section (figure g)
Sample: GMA PT 5 PNA 6.10
132 m   3°46'32.963" S 80°20'46.139" W
Observations: *Leucoagaricus* (Singer 1975:445): conforms to *Leucoagaricus rubrotinctus* (Peck) Singer 1948, with spores (6-9)x(4-5)μm. Also very similar to slightly bigger spored GMA PT 3 PSA 2.26 from Apurimac (P94b) and GMA PT 2 PSA 6.3A from Cusco (P94c). Regionally reported from Costa Rica (Mata et al., 2003), and North-Eastern Brasil (Neves et al., 2013:38).   Only old specimens were found. In Peru a viscous red *L. rubrotinotus* was also reported form Madre de Dios, Peru (Alvarez Loayza et al., 2014). See also:
https://www.mushroomexpert.com/leucoagaricus_rubrotinctus.htm

In Melzer's

1 unit = 3.75 μm

P94b
(2)

*cf. **Leucoagaricus rubrotinotus*** Spores: x̄ (9.5x 4.4)μm (Qm=2.1)

Southern Andes: Apurimac, Cusco

in Melzer's

1 unit = 3.75 μm

Region: Abancay, Apurimac and Zurite, Anta, Cusco - Southern Andes (figure g)
Ecology: organic matter in cloud forests under *Polycarpus glomeratus* (figure e) *and Polylepis racemosas and Polylepis besseri (*figure f)
Macroscopic
Odor: agreeable
Color reaction: none observed
Pileus: 4-5cm, umbonate when mature, with lepiotoid pattern of stretched-out fibrils over a white background, evenly colored and dark at center, dry (figure d)
Stipe: 6cm white, smooth, with membranous annulus that changes to cream or brown with age, and with no volva
Gillas: white, free, close
Spore print: white (figure c: orange arrow)
Microscopic
Spores:(9.4x10.1)x(4.5-5.0)μm, x̄ (9.5x 4.4)μm (Qm=2.1) (n=5), elliptical-subfusiform, smooth, thick walled, hyaline (KOH), lightly pseudoamyloid(Melzer's), without germ-pore (figure d)
Samples:
Abancay, Apurimac*: GMA PT 3 PSA 2.26
3273m   13°35'32.076" S 72°52'53.886" W
Zurite, Cusco   GMA PT 2 PSA 6.3
3908 m   13°25'55.301" S 72°15'18.707" W
Observations: Tentatively *Leucoagaricus* (Singer 1975:445): by similarity of carpophore and spores and to P94a from Tumbes as *Leucoagaricus rubrotinotus*, spores (6-9 x 4-5)μm, although this sample's spores are slightly larger than upper limit. *See: https://www.mushroomexpert.com/leucoagaricus_rubrotinctus.html.* Confirmation of pileipellis and cresyl blue spore stain similarity is need.

cf. **Leucoagaricus sp.** *(rubrotinotus?)* Spores: x̄ (10.1x5.1)µm (Qm=2.0)
Southern Andes: Apurimac

© Peter Trutmann

d

in Melzer's

1 unit = 3.75 µm

Region: Abancay, Apurimac (figure f)
Ecology: organic forest litter *Podocarpus glomeratus* cloud forest (fig. e)
Macroscopic
Pileus: 2-3cm, umbonate, surface dry to slightly moist, at first uniformly brown then breaking into flat, radially arranged fibrils or scales which vary in color from light to dark brown, background white, and center remains smooth and darker (figures a, b),
Stipe: 4cm, white to cream, with an intermediate membranous annulus, and slightly bulbous base, without volva (figures a, c)
Gills: white, free, crowded (figure c)
Spore print: white
Microscopic
Spores: (8-11)x(5-5.5)µm, x̄ (10.1x5.1)µm (Qm=2.0) (n=5), fusiform, with snout like ends (black arrow) smooth, hyaline (KOH), pseudoamyloide (Melzer's) without germ-pore (figure d)
Sample: GMA PT 3 PSA 2.18
3229 m   13°35'50.334" S 72°52'42.258" W
Observations: *cf. Leucoagaricus rubrotinotus* (Singer 1975:445): through comparison with similar samples P94a and b. This sample has slightly larger spore size than and darker pileus than *L. rubrotinotus, but* brown capped varieties of have been reported .

© Peter Trutmann

Leucoagaricus sp.   Spores: (9.7x5.3)µm (Qm=1.8)
Southern Andes: Cusco

in Melzer's

1 unit = 3.75 µm

Region: Santiago, Cusco -Southern Andes (figure f)
Ecology: soil in moss underneath *Eucalyptus globulus* (figure e)
Macroscopic
Odor: N.E.
Color reaction: none
Pileus: 4-5cm, broadly convex to umbonate, white, smooth, dry, sometimes with shaggy edges (figure a,b)
Stipe: 5cm, white to brown with age, with apical membranous intermediate annulus darkening with age on the underside, central, without volva (figure a,b)
Gill: white, free, crowded (figure c)
Spore print: white
Microscopic
Spores: (9-11)x(4.5-6)µm, x̄ (9.7x5.3)µm (Qm=1.8) (n=5), elliptical, smooth hyaline (KOH), pseudoamyloid (Melzer's), with narrow germ-pore (figure d)
Sample: GMA PT 2 PSA 4.3
3062 m   13°31'42.456" S 71°44'23.04" W
Observations: *cf. Leucoagaricus* (Singer 1975:445): by carpophore *Agaricus*-like form, a spore size a ≤10 µm with germ-pore. Not found in available national and regional literature. Pileipellis anatomy is lacking.

Leucoagaricus sp.   Spores: (9.4x6.6)µm (Qm=1.4)
Southern Andes: Ayacucho

Region: Puquio, Ayacucho - Southern Andes (figure d)
Ecology: highland agricultural field producing herbs (figure d)
Macroscopic
Color reaction: none
Pileus: 10-12cm, plane-umbonate, white, smooth, dry (figure a)
Stipe: 5-10(1)cm, thick, white, hollow, central, smooth, with fragile membranous, non-movable annulus, without volva (figure b)
Gills: white, free, crowded (figure b)
Spore print: white
Microscopic
Spores: (8.2-10.5)x(5.5-7.5)µm, x̄ (9.4x6.6)µm (Qm=1.4) , (n=5), elliptical, smooth, hyaline(KOH), dextrose (Melzer's), with narrow germ-pore (figure c)
Sample: GMA PT 2 PSA 1.3
3169 m   14°41'10.835" S 74°7'1.056" W
Observations: cf. *Leucoagaricus* (Singer 1975:445): has *Agaricus like* carpophore fleshy shape, smooth, pseudoamyloid spores with germ-pores. Similar to *L. leucothites*

in Melzer's

1 unit = 3.75 µm

cf. **Leucoagaricus sp.**   Spores: (6.7x4.9)µm (Qm= 1.3)
Central Andes: Ancash

Region: Carhuaz, Ancash - Central Andes (figure f)
Ecology: organic debris below unidentified native trees by river near Chaucayan (figure e)
Macroscopic
Odor: strong but agreeable
Pileus: 3-5cm, umbonate-plane, fleshy, pinkish brown above white base, with darker colored center, dry, (figures a,b) with appendiculate edges (figures b,c)
Stipe: 4-7.5(1)cm, white, smooth, with central intermediate membranous annulus, bulbous base but without volva (figures b,c)
Gills: white- light pink, free, crowded (figure c)
Spore print: did not yield print
Microscopic
Spores: (6.0-7.0)x(4.5-5.6)µm, x̄ (6.7x4.9) µm(Qm=1.3), (n=5), subglobose, hyaline (KOH), dextrose (Melzer's), smooth, thick walled, without germ-pore (figure d)
Pileipellis: N.E.
Sample: GMA PT 3 PNA 9.22
1667 m   10°9'20.435" S 77°31'19.608" W
Observations: *cf. Leucoagaricus* (Singer 1975:445): although spores do not have germ-pores the carpophore form and its pinkish gills (suggesting pinkish spores) point to a *Leucoagaricus* rather than *Lepiota*. Cresyl blue metachromatic stain of spores needs to be performed and pileipellis tested.

in Melzer's

1 unit = 3.75 µm

cf. Leucoagaricus
Southern Andes: Cusco

Spores: x̄ (7.6x4.3)μm (Qm=1.8)

© Peter Trutmann

© Peter Trutmann

in Melzer's

1 unit = 3.75 μm

c

d

e

© Peter Trutmann

Region: Chinchero, Cusco (fugure g)
Ecología: organic matter under *Eucalyptus* stand (figure f)
Macroscopic
Odor: mild
Pileus: 3-5cm, umbonate, first uniformly brown then breaking into flat, radially arranged fibrils or scales which vary in color from light to dark brown, with center smooth and darker, dry, cap sometimes splitting (figure a)
Stipe: 5 cm, white, quite robust, thicker at bottom than top, with a fragile non mobile membranous annulus, without volva (figs. a.b)
Gills: white, free, close (figure b)
Spore print: none (assumed white)
Microscopic
Spores: (7-8)x(4-4.5)μm, x̄ (7.6x4.3)μm (Qm=1.8) (n=5), elliptical, smooth, hyaline (KOH), pseudoamyloid (Melzer's) with large germ-pore (figure c).
Trama: parallel (black arrow), lightly pseudoamyloid (figure e)
Cystidia: none observed
Basidia: c. (30x8)μm, with 4 spores (figure d: white arrow)
Sample: GMA PT 3 PSA 3.18
3850 m   13°23'25.535" S 72°2'31.709" W
Observations: temporally *Leucoagaricus* (Singer 1975:445,451): due to its *robust form and* truncated, smooth, pseudoamyloid spores less than 10μm with large germ-pores. No clamp connections found, and annulus poorly developed. Missing pileipellis information.

Leucocoprinus

P99   *Leucocorpinus sp.*   Spores: x̄ (6.9x4.1)μm (Qm=1.7)
Northern Andes: Tumbes

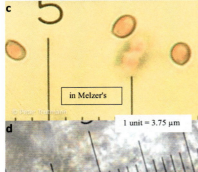

in Melzer's

1 unit = 3.75 μm

Region: Amotape, Tumbes, Tumbes - Northern Andes (figure f)
Ecology: organic material, close to river 'Bosque Tropical del Pacifico' with similarities to Amazonian forests (figure e)
Macroscopic
Odor: pleasant fungal
Pileus: 3.5-4.5cm, conical, white pulverulent base sometimes splitting below many small pieces of veil and a solid brown central calotte, eroded edges (figure a)
Stipe: 4-6cm, white turning brown, powdery and fibrous, with apical white annulus with brown lower side, without volva (figure b)
Gills: white, free, close (figure b)
Spore print: white (figure b: orange arrow)
Microscopic
Spores: (5. 6-7.5)x(3.8-4.5)μm, media (6.9x4.1))μm (Qm=1.7) (n=6), small, broadly elliptical to subglobose, small, hyaline(KOH), pseudoamyloid (Melzer's) without germ-pore (figure c)
Trama: N.E.
Cystidia: none found
Basidia: c. (18-20 x 4)μm
Pileipellis: a complex cellurar matrix: of inflated spherical cells - sphaerocysts (black arrow) -scalp section (figure d)
Sample: GMA PT 5 PNA 6.21
155 m   3°46'40.608" S 80°20'50.808" W
Observations: cf. *Leucocoprinus* (Singer 1975:445,456): due to coprinoid carpophore shape, what looks like a persistent, movable annulus and a complex pileipellis covered with sphaerocysts. However, spores are small and do not have germ-pores. Cresyl blue stain of spores is required. I am assuming a positive stain. If spores stain metachromatic then *Leucocoprinus,* if not then probably *Lepiota.* Sample was contaminated.

Macrolepiota

P100 ***Macrolepiota colombiana***   Spores: x̄ (13.0x8.9) μm (Q=1.5)
(2)   Southern and Northern Andes: Apurimac and Lambayeque

Region: Kaniaris, Ferreñafe, Lambayeque and Ampay, Abancay, Apurimac (figure i)
Ecology: organic matter and litter in northern relic Amazonian cloud forest (figure h) and
southern *Podocarpus glomeratus* cloud forest (figure j)
Macroscopic
Odor: pleasant
Pileus: 3.5-10 cm, convex to broadly convex, covered with brown veil when young, (figures
a,b), which breaks to form a mosaic of brown scales above the white epicutis below, with
dark calotte (figure c)
Stipe: 15-30 (1-1.5) cm, smooth, hard, brown, often eaten, with a moveable annulus, and
bulbous base, but no volva (figures, a,b)
Gills: white, free, crowded (figure d)
Spore print: did not yield print (presumed white)
Microscopic
Spores: (12.0-14.2)x(8.2-9.4) μm x̄ (13.0x8.9) μm (Qm=1.5) (n=5), elliptical, smooth, thick cell
wall, with broad germ-pore hyaline (KOH), pseudoamyloid (Melzer's) (figs.e,f)
Trama: not evaluated
Cystidia: abundant, crowded cheilocystidias (figure g: white arrow)
Basidia c. (40x8)μm with 4 spores (figure g: black arrow)
Pileipellis: not evaluated
Samples: Kaniaris, Lambayeque GMA PT 3 PNA 2.11
2719 m   6°3'41.916" S 79°15'7.283" W
            Ampay, Apurimac GMA PT 3 PSA 2.19
3277 m   13°35'47.849" S 72°52'44.91" W
Observations: conforms to *Macrolepiota colombiana* (Franco-Molano) (Franco-Molano,
1999) with spores 12-14(18) x 7-10(12) μm (Q = 1.56) Basidia (45-55(60) x 12-15 μm) and
similar cheilocystidia (n.b. the northern Peruvian specimen is twice maximum size described).

in Melzer's

1 unit = 3.75 μm

P101   *cf. Macrolepiota sp.*   Spores: x̄ (8.1x5.2)µm (Qm=1.5)
Northern Andes: Cajamarca

Region: Cajamarca - Northern Andes (figure d)
Ecology: organic material, in pasture close to a river (figure d)
Macroscopic
Odor: not tested (fresh) strong agreeable (dry)
Color reaction: none to reddish?
Pileus: 8cm, broadly conical, light brown over a white base (figure a)
Stipe: >11(0.5-1)cm, white, hollow, rigid and brittle when old, with a movable annulus (white upper surface, cream below), without volva (figure a,b)
Gills: white turning pink with age, free, close (figure b)
Spore print: not taken due to age of specimen (assumed white or pink)
Microscopic
Spores: (7.1-9.4)x(4.5-6.0)µm, x̄ (8.1x5.2)µm (Qm=1.5) (n=5), elliptical, smooth, with complex wall, hyaline (KOH), pseudoamyloide (Melzer's), with no or narrow pore, (black arrow)(figure c)
Trama: N.E.
Cistidia: N.E.
Basidia: N.E.
Pileipellis: N.E.
Sample: GMA PT 3 PNA 5.11
2715 m   7°13'59.165" S 78°16'5.73" W
Observations: tentatively *Macrolepiota* rather than *Leucoagaricus* (Singer 1975: 445): by carpophore shape and size with long (rather than short) stipe, movable annulus, red reaction although spores <10µm. It could also be a *Cholophyllum* sp. due to spore size and possible red color reaction. Presence of clamp connections needs checking as well as structure of pileipellis. One old specimen only.

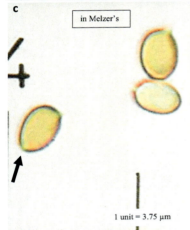

in Melzer's

1 unit = 3.75 µm

© Peter Trutmann

Family Amanitaceae

Catatrama

P102   *cf. **Catatrama** sp.*   Spores: x̄ (5.8x4.4)μm (Qm=1.3)
Southern Andes: Cusco

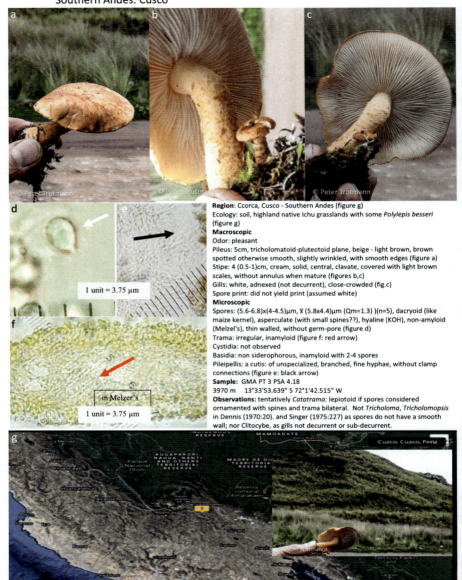

Region: Ccorca, Cusco - Southern Andes (figure g)
Ecology: soil, highland native Ichu grasslands with some *Polylepis besseri* (figure g)
Macroscopic
Odor: pleasant
Pileus: 5cm, tricholomatoid-pluteotoid plane, beige - light brown, brown spotted otherwise smooth, slightly wrinkled, with smooth edges (figure a)
Stipe: 4 (0.5-1)cm, cream, solid, central, clavate, covered with light brown scales, without annulus when mature (figures b,c)
Gills: white, adnexed (not decurrent), close-crowded (fig.c)
Spore print: did not yield print (assumed white)
Microscopic
Spores: (5.6-6.8)x(4-4.5)μm, x̄ (5.8x4.4)μm (Qm=1.3))(n=5), dacryoid (like maize kernel), asperculate (with small spines??), hyaline (KOH), non-amyloid (Melzel's), thin walled, without germ-pore (figure d)
Trama: irregular, inamyloid (figure f: red arrow)
Cystidia: not observed
Basidia: non siderophorous, inamyloid with 2-4 spores
Pileipellis: a cutis: of unspecialized, branched, fine hyphae, without clamp connections (figure e: black arrow)
Sample: GMA PT 3 PSA 4.18
3970 m   13°33'53.639" S 72°1'42.515" W
Observations: tentatively *Catatrama:* lepiotoid if spores considered ornamented with spines and trama bilateral. Not *Tricholoma, Tricholomopsis* in Dennis (1970:20). and Singer (1975:227) as spores do not have a smooth wall; nor Clitocybe, as gills not decurrent or sub-decurrent.

1 unit = 3.75 μm

in Melzer's

1 unit = 3.75 μm

Limacella

P103
(3)

cf. **Limacella sp.**
Southern Andes: Cusco

Spores: x̄ (9.5x5.7)μm (Qm=1.7)

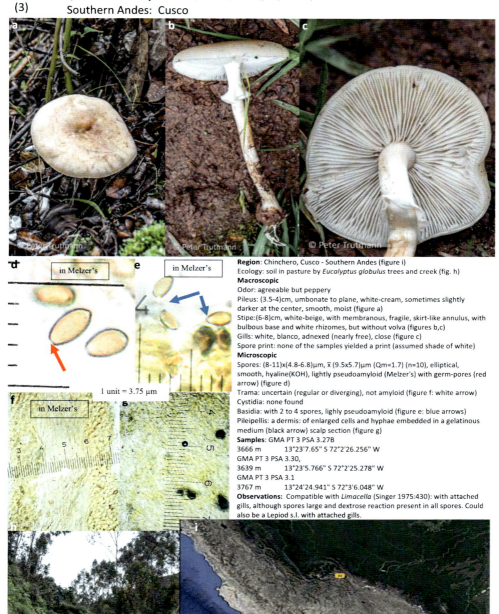

Region: Chinchero, Cusco - Southern Andes (figure i)
Ecology: soil in pasture by *Eucalyptus globulus* trees and creek (fig. h)
Macroscopic
Odor: agreeable but peppery
Pileus: (3.5-4)cm, umbonate to plane, white-cream, sometimes slightly darker at the center, smooth, moist (figure a)
Stipe:(6-8)cm, white-beige, with membranous, fragile, skirt-like annulus, with bulbous base and white rhizomes, but without volva (figures b,c)
Gills: white, blanco, adnexed (nearly free), close (figure c)
Spore print: none of the samples yielded a print (assumed shade of white)
Microscopic
Spores: (8-11)x(4.8-6.8)μm, x̄ (9.5x5.7)μm (Qm=1.7) (n=10), elliptical, smooth, hyaline(KOH), lightly pseudoamyloid (Melzer's) with germ-pores (red arrow) (figure d)
Trama: uncertain (regular or diverging), not amyloid (figure f: white arrow)
Cystidia: none found
Basidia: with 2 to 4 spores, lighly pseudoamyloid (figure e: blue arrows)
Pileipellis: a dermis: of enlarged cells and hyphae embedded in a gelatinous medium (black arrow) scalp section (figure g)
Samples: GMA PT 3 PSA 3.27B
3666 m   13°23'7.65" S 72°2'26.256" W
GMA PT 3 PSA 3.30,
3639 m   13°23'5.766" S 72°2'25.278" W
GMA PT 3 PSA 3.1
3767 m   13°24'24.941" S 72°3'6.048" W
Observations: Compatible with *Limacella* (Singer 1975:430): with attached gills, although spores large and dextrose reaction present in all spores. Could also be a Lepiod s.l. with attached gills.

in Melzer's

in Melzer's

1 unit = 3.75 μm

in Melzer's

Saproamanita

cf. ***Saproamanita sp.***   Spores: x̄ (7.4x5.8) μm (Qm=1.3)

Northern Coastal Hills: Tumbes

Region: Tumbes, Tumbes - Northern coastal hills (figure j)
Ecology: soil on the side of a dirt road under native trees (figure i)
Macroscopic
Odor: strong fish-like
Pileus: (6-10)cm, pluteotoid to plane, dry, white with a mosaic of wart-like scales, turning from white to brown closer to the center, margin smooth (figures a,b)
Stipe: (5-11)cm, white floccose with small scales, with membranous, skirt type annulus, not or only slightly bulbous at base with no indication of volva (figures c,d)
Gills: white, free, broad, crowded (figure d)
Spore print: did not yield print (assumed some form of white or off white)
Microscopic
Spores: (7.1-7.5)x(5.6-6.8)μm x̄ (7.4x5.8)μm (Qm=1.3) (n=6), subglobose, smooth or very finely ornamented, hyaline to green (KOH), lightly amyloid (Melzer's) (figures f,g: yellow arrows), without germ-pore (figure e)
Trama: regular or bilateral, made of inflated cells, lightly pseudoamyloid (figure g)
Cystidia: not found
Basidia: c. (15-20)x(4-8)μm , with 4 esporas (figure f: yellow arrow)
Pileipellis: a dermis: of inflated cells or shortened hyphae - scalp section (figure h)
Sample: GMA PT 5 PNA 5.8A
70 m   3°45'25.457" S 80°23'28.487" W
Observations: Tentatively *Saproamanita:* has all but a clear volva and universal veil characteristics of *Amanita:* including subglobose spores, lightly amyloid (when margin of pileus smooth) (Singer 1975:420). Extra-limited species also have amyloid spores according to Dennis (Dennis, 1970:50).

Family Bolbitiaceae

Bolbitius

P105   ***Bolbitius sp.***   Spores: x̄ (10.1x5.1)μm (Qm=2) (n=5)
Southern Andes: Cusco

Region: Chinchero, Cusco (figure i)
Ecology: soil in moss and grass in Eucalyptus plantation (fig. h)
Macroscopic
Odor and taste: chemical and peppery
Pileus: 2-4cm, umbonate and when older upturned, light - chocolate brown, often with cream margins, finely ridged, moist, shiny (fig.a,b)
Stipe: 4-5cm, smooth, white, without annulus or volva, easily separated from cap (figure c)
Gills: brown, adnate, crowded (figure c)
Spore print: chocolate brown (figure c: orange arrow)
Microscopic
Spores: (7.9-11.2)x(4.1-5.6)μm, x̄ (10.1x5.1)μm (Qm=2) (n=5), elliptical, dark brown (KOH), smooth, with germ-pore (figure d)
Trama: irregular (figure fL white arrow)
Cystidia: not found
Basidia: c. (30x 7.5)μm with 2 or more sterigmata (figure e: red arrow)
Pileipellis: a cellular dermis (cystodermis or hymenidermis) of enlarged, slightly pigmented cells (figure g:black arrow) - scalp section.
Sample: GMA PT 3 PSA 3.9
3823 m   13°23'27.024" S 72°2'40.32" W
Observation: *Bolbitius*: has cellular dermis and brown spore print.

1 unit = 3.75 μm

THE MACROFUNGI OF ANDEAN PERU   Part 1

Bolbitius sp.
Northern Andes: Cajamarca

Spores: x̄ (11.5x7.5) µm (Qm=1.5)

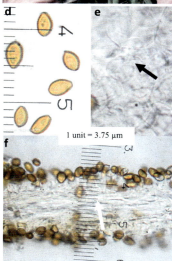

1 unit = 3.75 µm

Region: Pucara, Jaen, Cajamaraca - Northern Andes (figure h)
Ecology: soil and orgnic material close to river (figure g)
Macroscopic
Odor: strong
Pileus: 3-4.5cm, plane to umbonate, white with brown calotte, smooth, dry,dull (figure a)
Stipe: 6-9cm, white, central, fragile, without annulus or volva, base with rhizomes (figures b,c)
Gills: brown, emarginate, crowded (figure c)
Spore print: brown (figure c: orange arrow)
Microscopic
Spores: (8.2-13.1)x(7.1-7.9)µm, x̄ (11.5x7.5) µm (Qm=1.5) (n=5), ellipsoid, brown (KOH), smooth with germ-pore (figure d)
Trama: parallel (figure f: white arrow)
Cystidia: not found
Basidia: c. (20-25) x(6)µm with 2-4 spores (figure f)
Pileipellis: a cellular dermis: cellular cystoderma or hymeniderm (fig. e: black arrow)
Sample: GMA PT 3 PNA 2.33
 GMA PT 3 PNA 2.31
1135 m   6°0'31" S 79°11'49" W
Observations: *Bolbitius*: from spore characteristics and cellular pileipellis. Not found in local or regional literature (Mata et al 2006, Alvarez et al 2014, Cardenas 2019) or regionally (Dennis 1970:67, Franco-Molano et al 2005)

P107   *cf.* **Bolbitius sp.**   Spores: x̄ (12.4x8.0)μm (Qm=1.5)
(2)   Northern Andes: Lambayeque

1 unit = 3.75 μm

Region: Ferreñafe, Lambayeque - Northern Andes (figure g)
Ecology: soil, on the side of the road (figure f)
Macroscopic
Odor: not noteworthy
Pileus: 4-5cm, plane to upterned when mature, white to cream, smooth, moist,silky (figures a,b)
Stipe: 8(0.4)cm, white-cream, central, smooth, stipe easly separable from cap without annulus or volva, with mycelium in inferior part (figure a,b,c)
Gills: light brown, emarginate?, crowded (figures b,c)
Spore print: brown
Microscopic
Spores: (10.5-14.2)x(7.5-9.4)μm, x̄ (12.4x8.0)μm (Qm=1.5) (n=5), ellipsoide, brown(KOH),smooth, with germ-pore (figure d:black arrow)
Trama: N.E.
Cystidia: N.E.
Basidia: N.E.
Pileipellis: cellular dermis (cystoderm or hymeniderm) - scalp section (figure e)
Samples: GMA PT 3 PNA 2.2
1317 m   6°1'28.488" S 79°12'13.944" W
Observations: *Bolbitius* (Singer 1975:513-4): by spore characteristics, cellular pileipellis and spore print color.

Bolbitius titubans

Spores: x̄ (11.8x7.5) (Qm=1.6)

Southern and Central Andes: Cerro de Pasco, Cusco, Lima

© Peter Trutmann

© Peter Trutmann

1 unit = 2.5µm

1 unit = 1 um

10 µm

10 µm

10 µm

10 µm

Region: Canchis, Tipón, Cusco, la Quinoa, Cerro de Pasco, Canta, Lima (fig.l)
Ecology soil in agro-pastures (figures j,k)

Macroscopic
Pileus: 2-3cm, convex to conical, white to yellow-cream, often darker at center, fragile, finely corrugated (striated) plicate-sulcate, silky, glutinous (fig. a,b,m)
Stipe: 3-5cm, white to yellow, fragile, hollow, pulverulent, without annulus (fig. a)
Gills: rust brown, adnexed, close
Spore print: brown-rust brown (figure c: orange arrow)

Microscopic
Spores: (11-12) x (7-8) µm, x̄ (11.7x7.5) (Qm=1.6) (n=16), elliptical, brown to redish (KOH), smooth, with germ-pore (figure d)
Cystidia: brachybasidioles? (fig.h:white arrow)
Trama: appears regular (figure g:: red arrow)
Basidia:, c. (17x7)µm, forma cylindrical with 4 spores, (fig. i:black arrow)
Pileipellis: cellular dermis in gelatinous medium? (fig.e:yellow arrow) with pileicystidia like protrusions (fig. f: blue arrows)

Samples: Canchis, Cusco: GMA PT 1 PSA 7.1B
3528 m   14°10'24.071" S 71°22'13.578" W
La Quinoa, Cerro de Pasco: GMA PT 7 PCA 4.13
3741 m   10°38'17.706" S 76°10'16.847" W
Canta, Lima: GMA PT 7 PCA 1.2
                        GMA PT 7 PCA 1.2BF (
3337 m   11°27'16.115" S 76°35'50.706" W
Tipán, Choquepata, Cusco: GMA PT 1 PSA 5.14
3528 m   14°10'24.071" S 71°22'13.578" W

Observation: *Bolbitius titubans Bulliard, 1789) Fries, 1838*
(see.mushroomexpert.com/bolbitius_titubans)

GMA PT 7 PCA 1.2BF

Google Earth

Conocybe

Conocybe sp.   Spores: x̄ (15.5x9.0)µm (Qm=1.7)
Southern Andes: Cusco

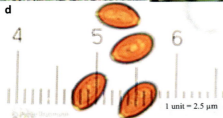

1 unit = 2.5 µm

Region: Quispichanchi, Cusco - Southern Andes (fig. h)
Ecology: soil in agro-pasture with *Pinus radiata* (figure g)
Macroscopic
Odor: N.E.
Pileus: 2-3cm, conical to plane and upturned when old, white-cream, dry, lightly corrugate and lightly circularly wrinkled and smooth (figures a,b)
Stipe: 4-5cm, cream, fragile, smooth changing to pruimose close to cap, dry, dull, without annulus or volva, central (figure c)
Gills: orange-brown, emarginate to adnexed, subdistant, wide (figure c)
Spore print: light ochre brown
Microscopic
Spores: (15-16.3)x(8.8-10)µm, x̄ (15.5x9.0)µm (Qm=1.7) (n=5), elliptical, smooth, red-brown with germ-pore (figure d)
Pileipellis: a dermis: of large globose cells: hymenidermis - scalp cut (figures e,f: black arrows)
Sample: GMA PT 1 PSA 5.2
3143 m   13°34'48.648" S 71°48'48.695" W
Observations *Conocybe* (Singer 1975: 513-14): by spore characteristics, carpophore conical shape and cellular epicutis.

1 unit = 2.5 µm

Conocybe sp.   Spores: x̄ (13.5x6.7)μm (Qm=2.0)
Southern Andes: Puno

Region: Putina, Puno - Southern Andes (figure g)
Ecology: soil in grass under *Polylepis besseri stand* (figure f)
Macroscopic
Pileus: 2cm, conical, orange brown, lightly rugulose and wrinkled, otherwise smooth, shiny, dry-glutenous, margin hygrophanous with age, silky, dry (figure a)
Stipe: 4cm, white, fragile, hollow, smooth, central, lacking annulus (figure a)
Gills: orange-brown, emarginate, close, thick (figure b)
Spores: rusty brown
Microscopic
Spores: (11.3-15)x(5-9)μm, x̄ (13.5x6.7)μm (Qm=2.0) (n=5) ellipsoid, reddish brown, smooth, unclear about presence of germ-pores (fig. c)
Cystidia: not found
Basidia: pyriform, with 2 or more sterigmata (figure d: white arrow)
Pileipellis: a dermis: a cellular, cysto or hymenidermis -scalp section (figure e: black arrow)
Sample: GMA PT 1 PSA 11.9
3874 m   14°54'39.27" S 69°52'38.051" W
Observations: *Conocybe* (Singer 1975: 413-14): carpophore conical, spore color and cellular epidermis

1 unit = 2.5 μm

P111   ***Conocybe sp.***   Spores: x̄ (15.7x9.2) µm (Qm=1.7)
Southern Andes: Lima

Region: Canta, Lima - Southern Andes (figure f)
Ecology: soil in overgrazed agro-pasture (figure f)
Macroscopic
Odor: N.E.
Pileus: 1.5cm, narrowly convex to conical, fragile, light
yellow, smooth, dry-glutinous, shiny, with crisped margin
(fig. a)
Stipe: 5cm, cream, fragile, hollow, pruinose, without
annulus (figures, a,b)
Gills: orange brown, emarginate, semi distant (figure d)
Spore print: did not yield print
Microscopic
Spores: (13.3-19.0)x(7.6-10.7)µm x̄ (15.7x9.2) µm (Qm=1.7)
(n=5), elliptical, orange brown (KOH), smooth, with germ-
pore (figure c)
Trama: parallel rather than irregular(figure e: white arrow)
Cystidia: none observed
Basidia: c.(15x7.5)µm, piriform with 2-4 sterigmata, (figure
d:black arrow)
Pileipellis: a cellular dermis
Sample: GMA PT 7 PCA 1.7
3426 m   11°27'20.244" S 76°35'44.045" W
Observations: *Conocybe* (see Singer 1975:514).

1 unit = 2.5 µm

Conocybe sp.

Spores: x̄ (13.2x8.7)µm (Qm=1.5)

Southern Andes: Arequipa, Puno

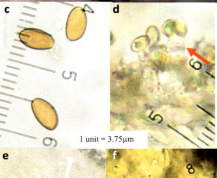

1 unit = 3.75µm

Region: Putina, Puno and Arequipa, Arequipa (figure h)
Ecology: soil, native highland ichu grasslands (figure g)
Macroscopic
Odor: N.E.:
Pileus: 2cm, conical, orange-cream, smooth, dry, dull (figure a)
Stipe: 4cm, cream, fragile, smooth, central, without annulus (figures a,b)
Gills: light orange-cream, attached, close (figure b)
Spore print: brown
Microscopic
Spores: (12.5-13.7)x(7.5-10.0)µm, x̄ (13.2x8.7)µm (Qm=1.5) (n=5) elliptical, smooth, ocre-brown, with germ-pore (figure c)
Trama: irregular?, of subglobose cells (figure e: white arrow)
Cystidia: not found
Basidia: with 2 or more spores (figure d: red arrow)
Pileipellis: a dermis: cellular in a gelatinous medium - scalp section (figure f: black arrow)
Sample: GMA PT 1 PSA 11.4B
3904 m   14,54.6079S 69,52.6193W
GMA PT 3 PSA 8.5BF
3918 m   16°3'25.248" S 71°21'20.592" W
Observations: *Conocybe* (see Singer 1975:514): due to conical cap shape, spore color and shape and cellular epidermis

Conocybe sp. Spores: x̄ (14.4x8.0)µm (Qm=1.8)

Southern Andes: Cusco

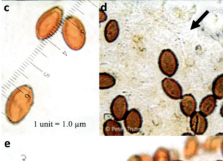

1 unit = 1.0 µm

1 unit = 2.5 µm

Region: Qorikancha, Cusco, Cusco - Southern Andes (figure g)
Ecology: soil in lawn (figure f)
Macroscopic
Odor: N.E.
Pileus: 1-2cm, campanulate, orange-cream, smooth, dry (fig. a)
Stipe: 4-5cm, cream, smooth, with vertical ridges, central, no annulus (figures, a,b)
Gills: orange, attached (adnexed?), sub-distant (figure b)
Spore print: rust brown
Microscopic
Spores: (13-15)x(8)µm, x̄ (14.4x8)µm (Qm=1.8) (n=5), elliptical, orange, ornamented, thick walled with germ-pore (figures c,d)
Trama: unclear (regular?) (figure e: white arrow)
Cystidia: not found
Bacidia: not found
Pileipellis: a collapsed dermis made of square to globular cells: (figure d: black arrow)
Sample:
GMA PT 1 PSA 5.1,
GMA PT 2 PSA 7.1
3138 m   13°34'48.929" S 71°48'48.971" W
Observations: *Conocybe* (see Singer 1975:514): carpophore form, spore print, spores with germ-pores and cellular epidermis. Sample in poor state.

Conocybe sp.

Spores: x̄ (11.4x6.2) μm (Qm=1.8)

Central Andes: Cerro de Pasco

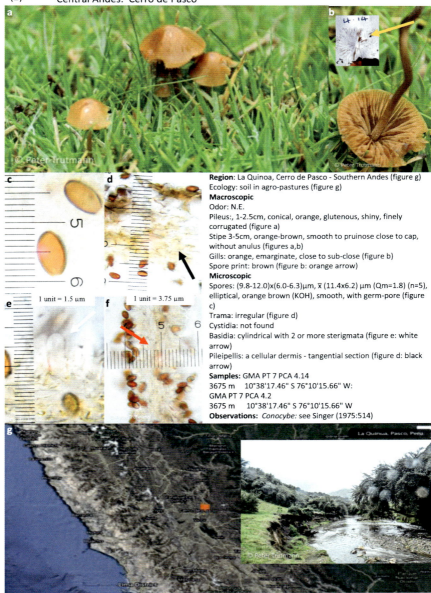

© Peter Trutmann

© Peter Trutmann

1 unit = 1.5 μm

1 unit = 3.75 μm

Region: La Quinoa, Cerro de Pasco - Southern Andes (figure g)
Ecology: soil in agro-pastures (figure g)
Macroscopic
Odor: N.E.
Pileus:, 1-2.5cm, conical, orange, glutenous, shiny, finely corrugated (figure a)
Stipe 3-5cm, orange-brown, smooth to pruinose close to cap, without anulus (figures a,b)
Gills: orange, emarginate, close to sub-close (figure b)
Spore print: brown (figure b: orange arrow)
Microscopic
Spores: (9.8-12.0)x(6.0-6.3)μm, x̄ (11.4x6.2) μm (Qm=1.8) (n=5), elliptical, orange brown (KOH), smooth, with germ-pore (figure c)
Trama: irregular (figure d)
Cystidia: not found
Basidia: cylindrical with 2 or more sterigmata (figure e: white arrow)
Pileipellis: a cellular dermis - tangential section (figure d: black arrow)
Samples: GMA PT 7 PCA 4.14
3675 m   10°38'17.46" S 76°10'15.66" W:
GMA PT 7 PCA 4.2
3675 m   10°38'17.46" S 76°10'15.66" W
Observations: *Conocybe:* see Singer (1975:514)

P115
(2)
Conocybe sp
Central Andes: Ancash
Spore: x̄ (12.1x8.6) µm (Q=1.4)

Region: Yungay, Ancash - Central Andes (figure f)
Ecology: soil in agro-pastures (figure f)
Macroscopic
Pileus: 2cm, conical, slightly orange when young to white to cream with an orange calotte when older, smooth, not hygrophanous, with smooth to eroded margin , dry, dull (figures a,b)
Stipe: 8cm, white, fragile, smooth, with fine vertical ridges close to cap, without annulus (figures a, b,c)
Gills: light orange-cream, emarginate, close
Spore print: did not yield print
Microscopic
Spores: (7.4-15)x(7.0-9.5)µm, x̄ (12.1x8.6) µm (Qm=1.4) (n=5), ovate to elongated rhomboid, orange-brown (KOH), smooth, with germ-pore (figure d)
Trama: N.E.
Cystidia: N.E.
Basidia: N.E.
Pileipellis: a cellular dermis - scalp sect. (figure e: black arrow)
Sample: GMA PT 3 PCA 8.5
2508 m   9°13'9.179" S 77°41'14.885" W
Yungay, Ancash: GMA PT 3 PNA 8.8, 2512 m   9°13'8.748" S 77°41'11.597" W
Observations: *Conocybe* (see Singer 1975: 514):

1 unit = 3.75 µm

136   THE MACROFUNGI OF ANDEAN PERU Part 1

Panaeolina sp.

Spores: x̄ (17.6x9.6)µm (Qm=1.9)

Southern Andes: Arequipa and Puno

1 unit = 2.5 µm

Region: Vacas, Arequipa and Huancané, Puno (fig. e)
Ecología: soil, native puna grassland (figure d)
Macroscopic:
Pileus: 1-3cm, narrowly convex to parabolic, gold -brown, shiny, dry, smooth (figure a)
Stipe: 4-5cm, brown, fibrous, lightly farinose by cap, without annulus, central (figure a)
Gills: brown, unevenly speckled, adnexed, close (figure a)
Spore print: dark purple
Microscopic
Spores: (13-22)x(8-12)µm, x̄ (17.6x9.6)µm (Qm=1.9) (n=17), elliptical, red-brown (KOH), ornamented, with germ-pore (fig.b: black arrows)
Trama: parallel
Cystidia: capitulate pleurocystidia?
Basidia: N.E.
Epicutis: a dermis of globose cells- scalp section (figure c)
Samples:
Vacas, Arequipa: GMA PT 1 PSA 12.3
3962 m   16°4'32.976" S 71°22'52.47" W
                    GMA PT 1 PSA 12.4
3962 m   16°4'32.976" S 71°22'52.47" W
Huancané, Puno: GMA PT 1 PSA 10.7
3883 m   15°10'34.595" S 69°50'41.532" W
Observations: *Panaeolina*: (Singer 1975:506): cellular dermis, brown ornamented spores with germ-pores. Tentatively separated from other Panaeolinas by shiny golden-brown cap and large spores. Found on mountain slope and sandy soil in native grassland near water. *Panaeolina* is synonymous with *Panaeolus*.

Panaeolina sp. Spores: x̄ (16.8x8.0) μm (Q=2.1)
Central Andes: Ancash

1 unit = 3.75 μm

4. 5

1 unit = 18.7 μm

Region: Huaraz, Ancash - Central Andes (figure g)
Ecology: on cattle dung, in highland pastures (figure f)
Macroscopic:
Pileus: 2-3cm, narrowly convex to conical, brown, smoothy, moist, hygrophanous, often with dark necrotic depressions (figures, a,b)
Stipe: 4-5cm, brown, lighter and farinose close to the cap, hollow, without annulus (figure c)
Gills: brown grey, adnexed?, close (figure c)
Spore print: dark purple (figure c: orange arrow)
Microscopic
Spores: (15-18)x(7.5-8.2)μm, x̄ (16.8x8.0) μm (Q=2.1) (n=5), elliptical, dark brown (KOH), ornamented, with germ-pore (figure d)
Trama: N.E.
Cystidia: N.E.
Basidia: N.E.
Pileipellis: a dermis- of globose cells - scalp sect. (figure e: black arrow)
Sample: GMA PT 3 PNA 7.1
4044 m   9°33'0.143" S 77°38'6.066" W
Observations: *Panaeolina* (Singer 1975:506): tentatively separated from other *Panaeolina* specimens by moist brown cap on dung. *Panaeolina* is synonymous with *Panaeolus*.

P122 **_Panaeolina sp._** Spores: x̄ (14.7x8.5)µm (Q=1.7)
Central Andes: Lima

Region: Canta, Lima - Central Andes (figure i)
Ecology: soil, in highland agro-pasture (figure h)
Macroscopic
Pileus: 0.3-1cm, narrowly parabolic to narrowly convex, cream, smooth, margin turned inwards (figures, a,b,c)
Stipe: 2-5cm, long, white-cream, white farinose close to cap, without annulus, central (figure c)
Gills: dark grey with white margin, adnexed, close (figure c)
Spore print: dark purple-black (figure c: orange arrow)
Microscopic
Spores: (13.5-15.0)x(7.5-9.0)µm, x̄ (14.7x8.5) µm (Q=1.7) (n=5), elliptical, dark brown (KOH), ornamented, with germ pore (figure d)
Trama; regular (figure f: red arrow)
Cystidia: pleurocystidia-like structures with pigmented content c.(22x7.5)µm (figure f: white arrow)
Basidia: c. (22.5x11)µm with 4 sterigmata (figure g: green arrow)
Pileipellis: a dermis of globose cells - radial section (figure e: black arrow)
Sample: GMA PT 7 PCA 1.8
3418 m   11°27'20.796" S 76°35'44.118" W
Observations: _Panaeolina (Singer 1975:506):_ tentatively separated by long stipe, cap size, form and pigmented cystidia.

Panaeolina sp.   Spores: x̄ (14.4x8.0)μm (Qm=1.8)
Northern Andes: Cajamarca

Region: Cerro el Castillo, Cajamarca - Northern Andes (figure f)
Ecology: soil in grass of highland agro-pasture (figure f)
Macroscopic
Odor: peppery
Pileus: 1.5-2.5cm, convex, light brown, dull, dry, hygrophanous (figures a, b)
Stipe: 4-10cm, white-cream, long, narrow, smooth, farinose close to cap, vertically striate, without annulus, central (figures b,c)
Gills: grey -brown, with white edge, emarginate-adnate, close (figure c)
Spore print: purple-brown
Microscopic
Spores: (12.0-15.7)x(7.5-8.2)μm, x̄ (14.4x8.0)μm (Qm=1.8) (n=5), elliptical, brown (KOH), ornamented, with germ-pore (figure d)
Trama: N.E.
Cystidia: N.E.
Basidia: N.E.
Pileipellis: a dermis: of globose cells (hymenidermis or cystodermis) - scalp section (figure e: black arrow)
Sample:
GMA PT 3 PNA 5.9
2745 m   7°13'34.397" S 78°16'43.116" W
Observations: _Panaeolina (Singer 1975:.506):_ tentatively separated from other Panaeolinas by carpophore shape and long stipe. _Panaeolina_ is synonymous with _Panaeolus._

1 unit = 3.75 μm

Pholiotina

*cf. **Pholiotina sp** .*   Spores: x̄ (13.2x6.6) μm (Qm=2)
Central Andes: Junín

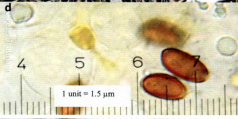

1 unit = 1.5 μm

1 unit = 3.75 μm

Región: Junín, Junín - Central Andes (figure f)
Ecology: soil in Ichu close to lago Junín (figure f)
Macroscopic
Pileus: 1-1.5cm, conical, red-brown with darker center
calotte, smooth, moist to viscous (figures a,b)
Stipe: 4 cm, brown with white powdery surface, without
annulus, central (figure c)
Gills: cream, with serrated edges, emarginate, close (fig. c)
Spore print: did not yield print
Microscopic
Spores: (12.0-14.3)x(6.0-7.5)μm, x̄ (13.2x6.6) μm (Qm=2)
(n=5), elliptical to oblong, brown (KOH), smooth, with
germ-pore (figure d)
Trama: parallel rather than irregular (figure e: red arrow)
Cystidia: none found
Basidia: c.(18x8)μm with 4 spores (figure e: white arrow)
Pileipellis: a dermis: of globular cells - scalp sect. (figure d)
Sample: GMA PT 7 PCA 6.2
4115 m   11°8'11.088" S 75°59'29.915" W
Observations: *Pholiotina* (Singer 1975: 518): like *Conocybe*,
but separated by having a parallel rather than irregular
trama

Family Clavariaceae

© Peter Trutmann

Clavaria

Clavaria sp.   Spores: x̄ (6.3x5.9)μm (Q=1.1)
Northern Andes: Cajamarca

Region: Colasay, Jaen, Cajamarca - Northern Andes (figure e)
Ecology: in organic material in relic Amazonian high selva cloud forest of Colasay (figure e)
Macroscopic
Fruiting body: (2.5-3.5)cm high (0.1-0.2)cm wide, delicate, unbranched, cylindrical with pointed ends (figures a,b)
Microscopic
Spores: (5.6-6.8)x(5.6-6.4)μm, x̄ (6.3x5.9)μm (Qm=1.1) (n=4) globose, hyaline (KOH), smooth without germ-pore (figure c,d: black arrows)
Trama: N.E.
Cystidia: not found
Basidia: (38-40)x(7-8)μm, non septate, irregularly shaped, hyaline (KOH) (figure d: red arrow)
Sample: GMA PT 5 PNA 3.9B
2435m   5°58'11.29" S 79°6'17.64" W
Observations: *Clavaria* (rather than *Clavariopsis*) (Arora 1986:635): due to spherical rather than oval spores and color of spore surface as well as finer fruiting body. Macroscopic finer and more delicate than 3.9A and has pointed ends. Microscopic has larger spores. Different than 3.10A by shape, color, viscosity, and presence of different cystidia and basidia. For information on taxonomy see (Birkebak et al., 2013).

Clavulinopsis

P126   **Clavulinopsis** *cf..fusiformis*   Spores: no spores found
Northern Andes: Lambayeque

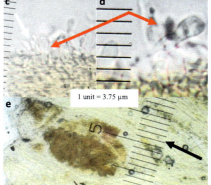

Region: Cañaris, Ferreñafe, Lambayeque (figure f)
Ecology: organic debris under relic Amazonian high selva cloud forest of Cañaris (figure f)
Macroscopic
Fruiting body: 1-2cm, cylindrical, with pointed extremities, not flat, orange, fragile, viscous, smooth (figures a,b)
Microscopic
Spores: none found
Internal tissue: Internal tissue: mycelium of partially inflated hyphae that appear monomitic. A section contained unidentified pigmented hyphae.
Outer spore baring surface: with larger clublike (basidia?) structures c.(34-38)x(7-8)μm (red arrows) and smaller non septate filiform structures (cystidia?) c.(15-20)x(3-4)μm (white arrow) (figures c,d)
Sample: GMA PT 6 PNA 3.10A
2721 m   6°3'41.76" S 79°15'7.902" W

Observations: *Clavulinopsis* (See Arora 1986:635). Peru, Madre de Dios (Alvarez Loayza et al., 2014; Cárdenas Medina et al., 2019), Colombia, Amazonas and Caquete Franco-Molano et al (2005:131). Known in Mexico in as 'escorbetas' by the Uitoto (Guzman, 1997). See also (Birkebak et al., 2013). Not a *Colacera* in the Dacrymycetaceae with branched basidias.found in the high selva of Cerro de Pasco (Salvador Montoya, 2009).

1 unit = 3.75 μm

Clavulinopsis sp. Spores: x̄ (4.5x4.2)μm (Q=1.1)
Northern Andes: Cajamarca

Region: Colasay, Jaen, Cajamarca - Northern Andes (figure e)
Ecology: in organic material in relic amazonian high selva cloud forest
of Colasay (figure e)
Macroscopic
Fruiting body: (2.5-3.5)cm high, (0.3-0.4)μm wide, yellow, cylindrical,
ends plane not pointed (figures a,b)
Microscopic
Spores: (4.2-4.9)x(4.2-4.5)μm, x̄ (4.5x4.2)μm (Qm=1.1) (n=5), appear
globose, hyaline (KOH), smooth, without germ-pore (fig. d:black arrow)
Cystidia: not found
Basidia: not separate nor branched c.(56)x(7-8) μm (red arrow) with
two or more sterigmata (figures c,d: white arrows)
Sample: GMA PT 5 PNA 3.9A
2435 m   5°58'11.29" S 79°6'17.64" W
Observations: _Clavulinopsis_ Arora (1986:635). Similar to, but different
from 3.10 morphology. It is more robust, yellow with flat and non-
pointed extremities, and microscopic differences. Not a _Colacera_ in the
Family Dacrymycetaceae found in the high jungle of Cerro de Pasco
(Salvador Montoya, 2009) having branched basidia. For information on
taxonomy and ecology see(Birkebak et al., 2013).

1 unit = 3.75 μm

cf. **Clavulinopsis**
Southern Andes: Cusco

Region: Zurite, Cusco - Southern Andes (figure c)
Ecology: organic material with moss in *Polylepis besseri* cloud forest (figure c)
Macroscopic
Fruiting body: (1-4)cm high, thin, cylindrical with pointed ends, yellow, unbranched, smooth (figures a,b)
Photographic sample:
GMA PT 2 PSA 6.16F
3997 m   13°25'48.911" S 72°15'23.982" W
Observations: probably *Clavariopsis*: with simple carpophores Arora (1989: 635).

P129F *Clavulinopsis ?*   Photographic record only
Southern Andes: Cusco

Region: Zurite, Anta, Cusco -Southern Andes (figure c)
Ecology: organic material with moss in *Polylepis besseri* cloud
forest (figure c)
Macroscopic
Fruiting body: (1-4)cm, pointed, cylindrical, long but robust,
growing in clumps, yellow, smooth (figure b)
Photographic sample: GMA PT 3 PSA 7.21D
4040m   13°25'46.998" S 72°15'24.527" W
Observations: possibly *Clavulinopsi*s or *Clavaria*: carpophores
simple or sparsely branched, not capitulated (Arora 1989: 635)

Indetermined coral fungi

P130F *Indetermined coral fungus (Clavaria sp?)*   Photographic record only
Southern Andes: Cusco

Region: Zurite, Cusco Southern Andes (figure c)
Ecology: organic material in cloud forest of *Polylepis besseri*
(figures a,c)
Macroscopic
Fruiting body: 1-2cm, coral like, stout, slightly branched,
white (figure b)
Spores: indetermined
Photograpic sample: GMA PT 2 PSA 6.12F
3920 m 13°25'55.02" S 72°15'18.179" W
Observations: Unidentifiable, but likely in Clavariaceae.

P131F
Indet. Coral fungus (Ramariopsis?) Photographic record only
Northern Andes: Lambayeque

© Peter Trutmann

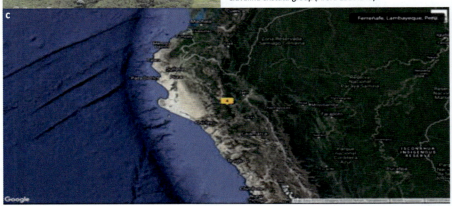

Region: Cañaris, Ferreñafe, Lambayeque (figure c)
Ecology: leaf litter, under relic Amazonian high selva cloud forest of Caniaris (figure b)
Macroscopic
Fruiting body: 1-2cm, coral like, delicate, occasionally forked,, white (figure a)
Photographic sample: GMA PT 5 PNA 2.28BF
2405 m   6°4'24" S 79°14'4" W
Observations: unidentifiable but appears a Clavariaceae and related to *Ramariopsis* or *Clavulina* being fragile, brittle, without elongated apices, without teeth or crown. It is not an *Artomyces* because the apices do not have a crown that was reported from Madre de Dios (Cárdenas Medina et al., 2019: no.418-420). (N.B. for lack of a crown in the example of Madre de Dios). The sample appears similar to the *Clavulina cristata* group (Arora 1986:641).

Family Cortinariaceae

© Peter Trutmann

Cortinarius

Cortinarius sp.   Spores:, x̄ (6.6x4.7)µm (Qm=1.4)
Central Andes: Cerro De Pasco

Region: Vicco, Cerro de Pasco Central Andes (figure k)
Ecology: soil in Ichu, puna grasslands close to Lake Junín (figure j)
Macroscopic
Pileus: 1cm, convex, fleshy, yellow cream, smooth, with in-turned margins (figures a,b)
Stipe: 3(1)cm, yellow-cream, with fine darker tissue on outside, and possibly very fine remnants of a slimy fibrous veil (figure b,c)
Gills: orange, adnexed, crowded (figure c)
Spore print: no
Microscopic
Spores (6.0-7.5)x(4.5-4.8)µm, x̄ (6.6x4.7)µm (Qm=1.4) (n=5), : amigdaliform, brown (KOH), lightly ornamented, without germ-pore (figure d)
Trama: parallel - regular (figure g: white arrow)
Cystidia: Looks like small cylindrical cleistocystidia c.(22.5-30)x (2-2.5)µm (figure i: yellow arrow) and fine pleurocystidia (figure h: red arrow)
Basidia: c. (25.5-27)x(6)µm with 4 spores (figure h)
Pileipellis: a dermis: (trichoderm?) of upright unspecialized hyphae (black arrow) above a cellular subcutis - scalp section (figure e) - radial section (figure f) (Largent et al., 1977) p.58
Sample: GMA PT 7 PCA 5.5
4145 m   10°49'27.774" S 76°12'11.04" W
Observations: *Cortinarius*: Dennis (1970:71), Singer (1975:587).

Indetermined Cortinariaceae

Region: Cañaris, Ferreñafe, Lambayeque - Northern Andes (figure g)
Ecology: wood, on dead branch of tree by side of the road. (figure g)
Macroscopic
Odor: no
Pileus: (1-2)cm, plane and upturned when older, red brown, thin, smooth, dry (figure a)
Stipe: (3-4)cm, brown-reddish, tough, central, without annulus of volva (figs.a,b)
Gills: orange, adnate, sub-close (figures c,d)
Spore print; did not yield print
Microscopic
Spores: not found
Trama: parallel and very lightly pseudoamyloid in Melzer's (fig. e:white arrow)
Cystidia: not found
Basidia: c.(38x7-8), upper part amyloid (figure d: red arrow)
Pileipellis: a cutis or trichoderm: of inflated mycelium and possibly palisade like subcutical cells- scalp section (figure f: black arrow)
Sample: GMA PT 6 PNA 2.2
1320 m   6°1'55.302" S 79°12'10.727" W
Observations: for its coloration placed in the Cortinaceae. similar to *Pyrrhoglossum* by gill color, hymenophore trama regular, absence of cystidia, absence of veil and epidermis with pigment.

1 unit = 3.75 μm

Family Crepidotaceae

© Peter Trutmann

Crepidotus

P134F   ***Crepidotus sp.***   Photographic record only
Southern Andes: Cusco

Region: Chinchero, Cusco - Southern Andes (figure d)
Ecology: on the trunk of a cut *Eucalyptus* tree (figures c,d)
Macroscopic
Pileus: 1cm, pleurotoid, cream, smooth, dry, with laciniate cracking of the surface (figure a)
Stipe: reduced, acentric (figure b)
Gills: white-cream, broad, uncertain attachment, distant (figure b)
Photographic sample only: GMA PT 3 PSA 3.10BF
3829 m   13°23'28.086" S 72°2'37.529" W
Observations: *Crepidotus:* due to form and color of gills as indicator of spore color, as well as its growth on wood (Arora 1989: 405).

Crepidotus brunswickianus (Speg.) Sacc.   *Spores:* x̄ (7.2x7.3)µm (Qm=1.0)
Central Andes: Ancash

Region: Yungay, Ancash - Central Andes (figure g)
Ecology: on a dead tree trunk in *Polylepis besseri* forest
dominated cloud forest (figure g)
Macroscopic
Pileus: (2-2.5)cm, pleurotoid - bracket like, kidney shaped, yellow
cream with orange patches, soft, smooth (figures a,b)
Stipe: rudimentary, off center (figure c)
Gills: white to cream (figure c)
Spore print: yellow brown (figure c: orange arrow)
Microscopic
Spores: (6.8-7.5)x(6.8-7.9)µm, x̄ (7.2x7.3)µm (Qm=1.0) (n=5),
globose to subglobose, light brown (KOH), ornamented, without
germ-pore (figure d)
Trama: N.E.
Cystidia: not found, only oil drops when checked in Melzer's
(figure f: black arrow)
Basidia: N.E.
Pileipellis: a cutis: as a carpet of mycelium - scalp section (figure
e: black arrow)
Sample: GMA PT 5 PNA 1.17B
3853 m   9°3'50.074" S 77°38'1.866" W
Observations: Like *Crepidotus brunswickianus (Speg.) Sacc,* with
oil globules in Melzer's. Reported in Paramos in Venezuela
Dennis (1970:75), from Chile and Argentina (Lazo, 2001:222)

1 unit = 3.75 µm

Crepidotus sp. Spores: c.3x3μm, x̄ 3x3μm (Qm=1.0)
Northern Andes: Lambayeque

Region: Cañaris, Ferreñafe, Lambayeque -Northern Andes (figure g)
Ecology: on bamboo, dead branch in relic high selva Amazonian cloud forest of Cañaris (figure g)
Macroscopic
Odor: mild
Pileus: (1-1.5)cm, campanulate - pleurotoid, light brown, smooth shiny, delicate, with semi upturned margins, slightly moist (figure a)
Stipe: reduced (0.3-0.5)cm, off-center
Gills: white to light brown, adnate, shallow, close to crowded (figures b,c)
Spore print: did not yield print
Microscopic
Spores: c. (3x3)μm, x̄ (3x3) (Qm=1) (n=1) sparce and very small, globose, hyaline (KOH), smooth without germ-pore (figure d: white arrow)
Trama: parallel (figure e: orange arrow)
Cystidia: not found
Basidia: c(24xx8)cm with at least 2 sterigmata (figure d: red arrow)
Pileipellis: a cutis of parallel hyphae- scalp section (figure f:black arrow)
Sample: GMA PT 3 PNA 2.11B
2707 m   6°3'42.293" S 79°15'7.703" W
Observation: *Crepidotus*: like *C. polypedidis (*Dennis 1970:75) that was recorded from the paramos Sierra de Santa Domingo (Venezuela). Its mature spores are globose (6-7μm). The species is only found on bamboo. A similar Crepidotus has not been reported from Peru.

162   THE MACROFUNGI OF ANDEAN PERU   Part 1

Crepidotus sp.   Photographic record only
Northern Andes: Lambayeque

Region: Cañaris, Ferreñafe, Lambayeque -Northern Andes (figure d)
Ecología: on bamboo, dead branch in relic high selva Amazonian cloud forest of
Cañaris (figure d)
Macroscopic
Pileus: (0.3-0.5)cm, pleurotoid, light brown, smooth, dry, dull (figure b)
stipe: reduced (0.1)cm, off-center (figures a,c)
Gills: white to cream, adnate, close
Spore color: assuming light brown due to color of older gills
Microscopic
Photographic sample only: GMA PT 5 PNA 2.26BF
2450 m   6°4'23.243" S 79°14'2.561" W
Observations: _Crepidotus_ assuming brown spores: and because of its
pleurotoid form, cream gills and lignicolous habit. Only found on bamboo and
has not been found in reports from other parts of Peru.

Simocybe

Simocybe sp.?   Spores: x̄ (9.6x5.2)µm (Qm=1.8)
Central Andes: Lima

Region: Canta, Lima - Central Andes (figure h)
Ecology: wood, on a dead branch of a bush in a degraded
environment (figures a, h)
Macroscopic
Pileus: (0.3-0.5)cm, bracket form, with slightly upturned edge, ,
white-cream, soft, delicate, slightly moist (figure b)
Stipe: reduced, off-center (figure c)
Gills: white-cream (figure c)
Spore print: did not yield print
Microscopic
Spores: (9.0-9.8)x(5.2)µm, x̄ (9.6x5.2)µm (Q=1.8) (n=5), elliptical,
light olive brown (KOH) ornamented, without germ-pore (figure d)
Trama: parallel (figure f: red arrow)
Cystidia: not found
Basidia: c. (22.5x6)µm (yellow arrow) with 4 spores (figure g)
Pileipellis: a dermis: of what appear subglobose cells in a
gelatinous subdermis (black arrow). The epidermis appears
colonized by a Deuteromycete with melanated, multiseptate
conidia (figure e)
Sample: GMA PT 7 PCA 1.13
3612 m   11°26'32.094" S 76°35'6.623" W
Observation: I'm tentatively placing it in *Simocybe* for its pileus
form, its elliptical and olive brown spores, and regular
hymenophore trama. Appears not reported from Peru

Family Entolomataceae

© Peter Trutmann

Clitopilus

Clitopilus *(syn. Rhodocybe)*   Spores: x̄ (6.2x 5.4)μm (Qm=1.1)
Southern Andes: Cusco

Region: Quispichanchi, Cusco - Southern Andes (figure i)
Ecology: soil in an agro-pasture bordered by pines (figure h)
Macroscopic
Odor: N.E.
Pileus: (3-5)cm, plane, cream, with darker and lighter radial zones, fleshy, smooth, dry, silky-dull (figure a)
Stipe: 3(1-1.5)cm, white, solid, acentric, smooth, without annulus (figure b)
Gills: white to cream, decurrent, crowded (figure c)
Spore print: cream (figure c: orange arrow)
Microscopic
Spores: (6-7)x(5-6)μm, x̄ (6.2x 5.4)μm (Qm=1.1) (n=5), radially (polar view) 5-6 angular (red arrow), longitudinally oval, with small vacuoles or inclusions, hyaline (KOH), inamyloid, smooth, without germ-pore (figure d)
Trama: unclear
Cystidia: not found
Basidia: long c. 25μm, with inclusions (figure e: blue arrow)
Pileipellis: a cutis: of dense repent narrow hyphae (figure g: black arrow)
Sample: GMA PT 1 PSA 5.5
3135m   13°34'48.815" S 71°48'48.893" W
Observations: *Rhodocybe*: by spore color, polar view angular spores, basidia with inclusions (siderophores) (see Singer 1975: 669). Syn. *Clitopilus* (Co-David et al., 2009):

1 unit = 1 μm

1 unit = 2.5 μm

1 unit = 2.5 μm

Rhodophana

Rhodophana sp.   Spores: x̄ (5.8x4.3)μm (Qm=1.3)

P140

Northern Andes: Lambayeque

© Peter Trutmann

© Peter Trutmann

© Peter Trutmann

1 unit = 3.75 μm

Region: Salas, Lambayeque - Northern Andes (figure i)

Ecology: in leaf debris under a large mother tree in small stand (fig. h)

Macroscopic

Odor: aromatic

Pileus: 5cm, plane to shallowly depressed, light brown with darked center, slightly undulated surface, smooth (fig. a)

Stipe: (5-6)cm, white, tough, centric, farinose, without annulus or volva but base with white mycelium on substrate (figures b,c)

Gills: white, adnate to sinuate, crowded (figure c)

Spore print: did not yield print

Microscopic

Spores: (4.5-6.7)x(3.7-5.2)μm x̄ (5.8x4.3)μm (Q=1.3) (n=5), echinulate warty ornamentation, greenish-hyaline (KOH, non-amyloid (Melzer's) without germ-pore (figures d,e: white arrow)

Cystidia: not found

Basidia: c. (25-28 x7-8)μm with 2 or more (figure f: red arrow)

Pileipellis: A cutis: prostrate fine hyphae above a cellular subdermis- scalp section (figure g: black arrow)

Sample: GMA PT 3 PNA 1.5

1122 m   6°13'30.641" S 79°31'39.803" W

Observations: *Rhodophana*: spores are ornamented, adorned with small bumps and gills are adnate or very short decurrent (Laessoe and Petersen, 2019:448). Gills eaten by insects.

Entoloma

sub-Genus Entoloma

P141   ***Entoloma sp.***   Spores:x̄ (9.0x7.5)μm (Qm=1.2)
(8)   Southern Andes: Cusco and Puno

Region: Putina, Puno; Cusco, Cusco - Southern Andes (figure f)
Ecology: soil, in native highland puna grasslands (figure e)
Macroscopic
Pileus: (3-4)cm, conical, grey-dark brown, often radially zoned,
smooth, moist, margins often lighter brown (figure a,b)
Stipe: 3cm, brown- farinose white, central, hollow without annulus
(figure b)
Gills: creamy pink, emarginate, close
Spore print: very light orange (figure b: orange arrow)
Microscopic
Spores: (7.5-11.0)x(5-9)μm, x̄ (9.0x7.5)μm (Qm=1.2) (n=10), (5-7)
angular, hyaline (KOH), non-amyloid (Melzer's), smooth (figures c,d)
Cystidia and Basidia: not found
Pileipellis: a cutis: a prostrate carpet of mycelium

Samples:
Putina, Puno : GMA PT 1 PSA 11.4,
GMA PT 1 PSA 11.5
GMA PT 1 PSA 11.6,
GMA PT 1 PSA 11.7,
GMA PT 1 PSA 11.10,
GMA PT 1 PSA 11.11
3858 m   14"59'24.245" S 69"50'29.687" W
GMA PT 2 PSA 9B3
3854m   14"58'31.8" S 69"50'41.879" W
Cusco, Cusco: GMA PT 3 PSA 4.12F
4011 m   13"33'55.224" S 72"1'39.515" W

Observations: *Entoloma:* carpophore shape, spore print and angular
spores. Samples in poor condition

1 unit = 2.5 μm

1 unit = 1 μm

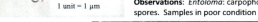

P142 **_Entoloma sp._** Sporess: x̄ (12.3x8.9)µm (Qm= 1.4)
Southern Andes: Cusco

Region: Cusco, Cusco - Southern Andes (figure g)
Ecology: soil in native Ichu puna grasslands (figure f)
Macroscopic
Pileus: (1-2)m, conical, brown with lighter margin, smooth, dry (figure a)
Stipe: (2.5-4)cm, brown, fibrous, smooth, no annulus, central (fig. b)
Gills: creamy pink, adnate, close (figure b)
Spore print: orange (figure b: orange arrow)
Microscopic
Spores: (11-14,5)x(8-9.5)µm, x̄ (12.3x8.9)µm (Qm= 1.4) (n=5), (6-9) angular, smooth, hyaline (KOH), inamyloid (Melzer's), without germ-pore (figure c)
Trama: parallel (figure e: white arrow)
Cystidia: not found
Basidia: c.(35 x7)µm, at least 2 spores (fig.d:black arrow)
Pileipellis: cutis : carpet of mycelium
Sample: GMA PT 3 PSA 4.17
3992 m   13°33'54.606" S 72°1'41.441" W
Observations: *Entoloma:* carpophore shape, color spore print, and (6-9) angular spores. Smaller than GMA PT 3 PSA 4.21 which has a corrugated cap and smaller spores

1 unit = 3.75 µm

Entoloma sp.
Southern Andes: Cusco

Spores: x̄ (11.0x8.2)µm (Qm=1.3)

Region: Cusco, Cusco - Southern Andes (figure i)
Ecology: soil in highland Ichu puna shrub-grassland (figure h)
Macroscopic
Odor: delicious
Pileus: (1.5-3.5)cm, conical, brown, darker at center, finely corrugate, dry (figures a,b)
Stipe: 10-15(0.5)cm, white to brown, filamentous, slightly farinose near gills, without annulus, central (figures b,c)
Gills: creamy pink, emarginate, close (figure c)
Spore print: dark orange (figure c: orange arrow)
Microscopic
Spores: (10.5-11,5)x (8-8.5)µm, x̄ (11.0x8.2)µm (Qm=1.3) (n=5), (6-9) angular, smooth, lightly cream (KOH), inamyloid (Melzer's), smooth, without germ-pore (figure d)
Trama: N.E.
Cystidia: not found
Basidia: c.(41x12)µm, with 2 sterigmata (figure e: white arrow)
Pileipellis: a dermis: of inflated hyphae (black arrow) covered with pigment - scalp section (figure g)
Sample:
Cusco, Cusco: GMA PT 3 PSA 4.21,
GMA PT 3 PSA 4.25F
3978 m   13°33'53.591" S 72°1'42.713" W
Anta, Cusco: GMA PT 3 PSA 5.4F
3775m   13°36'10.878" S 72°12'16.847" W
Observations: *Entoloma*: carpophore shape and color, spore print, and (6-0) angular spores. See also GMA PT 3 PSA 4.17.

1 unit = 3.75 µm

Entoloma sp.   Spores: x̄ (9.1x9.0)μm (Qm=1.0)
Southern Andes: Cusco

Region: Choquepata, Cusco - Southern Andes (figure h)
Ecology: organic matter, in grassland near Tipón (fig.h)
Macroscopic
Pileum: 3cm, conical, dark brown, finely striate, smooth, dry (figure a,b)
Stipe: 4 cm, dark brown, thin, smooth, central, without annulus (figures b,c)
Gills: creamy pink to white, emarginate-adnate, close
Spore print: did not yield print
Microscopic
Spores: (8-10)x(8-10)μm, x̄ (9.1x9.0)μm (Qm=1.0) (n=5), (5-6) angular, hyaline with green vacuole (KOH), non-amyloid (Melzer's), smooth, without germ-pore (figure d)
Trama: parallel (figure f: white arrow)
Cystidia: not found
Basidia. c.(35x12)μm, with 2 or more sterigmata (figure e: red arrow)
Pileipellis: a cutis or trichoderm: of enlarged parallel hyphae - scalp section (figure g: black arrow)
Sample: GMA PT 1 PSA 5.17
3349 m   13°34'26.747" S 71°46'57.485" W
Observations: *Entoloma:* differs from GMA PT 3 PSA 4.21 by more delicate stipe and spore size and shape.

1 unit = 2.5 μm

Entoloma sp
Southern Andes: Ayacucho

Spores: x̄ (9.7x6.2)μm (Qm=1.6)

a

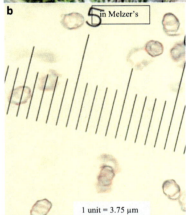

b

1 unit = 3.75 μm

Region: Lucanas, Ayacucho - Southern Andes (figure c)
Ecology: soil, in pasture with Lupinus shrubs (figure c)
Macroscopic
Pileus: (2-4)cm, conical, to broadly conical, dark brown, with radial zones (figure a)
Stipe: (3-4)cm, white, vertically lined, robust, central, without annulus (figure a)
Gills: cream, emarginate, close to crowded
Spore print: creamy pink (figure a: orange arrow)
Microscopic
Spores: (9.5-11)x(5.5-7.5)μm, x̄ (9.7x6.2)μm (Qm=1.6) (n=5), (6-11) angular, hyaline (KOH), non-amyloid (Melzer's), smooth, without germ-pore (figure b)
Cystidia and Basidia (not evaluated)
Pileipellis: a cutis (or trichdermis?): of parallel hyphae
Sample: GMA PT 2 PSA 2.3
4135 m 14°37'46.823" S 74°2'15.137" W
Observations: *Entoloma*: carpophore shape, spore print and spore shape. See Arora (1989: 242), Dennis (1970: 76) and Singer (1975:673)

c

Entoloma sp.
Central Andes: Junín

Spores: x̄ (10.8x7.4)μm (Qm=1.5)

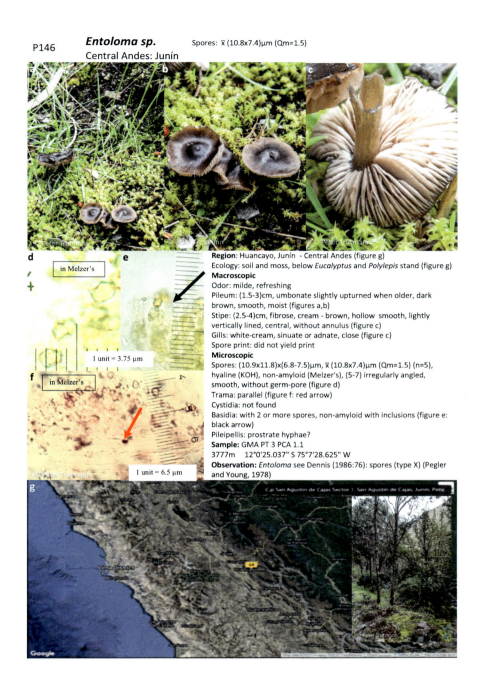

Region: Huancayo, Junín - Central Andes (figure g)
Ecology: soil and moss, below *Eucalyptus* and *Polylepis* stand (figure g)
Macroscopic
Odor: milde, refreshing
Pileum: (1.5-3)cm, umbonate slightly upturned when older, dark brown, smooth, moist (figures a,b)
Stipe: (2.5-4)cm, fibrose, cream - brown, hollow smooth, lightly vertically lined, central, without annulus (figure c)
Gills: white-cream, sinuate or adnate, close (figure c)
Spore print: did not yield print
Microscopic
Spores: (10.9x11.8)x(6.8-7.5)μm, x̄ (10.8x7.4)μm (Qm=1.5) (n=5), hyaline (KOH), non-amyloid (Melzer's), (5-7) irregularly angled, smooth, without germ-pore (figure d)
Trama: parallel (figure f: red arrow)
Cystidia: not found
Basidia: with 2 or more spores, non-amyloid with inclusions (figure e: black arrow)
Pileipellis: prostrate hyphae?
Sample: GMA PT 3 PCA 1.1
3777m   12°0'25.037" S 75°7'28.625" W
Observation: *Entoloma* see Dennis (1986:76): spores (type X) (Pegler and Young, 1978)

in Melzer's

1 unit = 3.75 μm

in Melzer's

1 unit = 6.5 μm

Entoloma sp.

Spores: x̄ (12.8x8.0)µm (Qm=1.5)

Central Andes: Ancash

1 unit = 3.75 µm

in Melzer's

1 unit = 6.5 µm

in Melzer's

1 unit = 6.5 µm

in Melzer's

1 unit = 3.75 µm

Region: Yungay, Ancash - Central Andes (figure h)
Ecology: soil and moss, in *Polylepis besseri* cloud forest (figure h)
Macroscopic
Odor:- no odor
Pileus: 2cm, broadly conical, brown narrowly striate, with shallow and deeper tares, smooth, slightly humid (figure a)
Stipe: 4cm, dark brown, slender, smooth, central, no annulus (fig.b)
Gills: white to pink, emarginate or adnexed, crowded (figure c)
Spore print: pink to orange
Microscopic
Spores: (9.5-16.8)x(7.0-13.1)µm, x̄ (12.8x8.0)µm (Qm=1.5) n=(20), irregular 5-7 angled, hyaline (KOH), lightly dextrose (Melzer's), without germ-pore (figure d)
Trama: parallel (figure f: white arrow)
Cystidia: not found
Basidia: c. (20x8)µm , 2 sterigmata, with lightly dextrose staining inclusions (Melzer's) (figure g: red arrow
Pileipellis: a cutis or trichoderm: of enlarged hyphae - scalp section (figure e: black arrow).
Sample:
GMA PT 3 PNA 8.12
GMA PT 5 PNA 1.6B
GMA PT 3 PNA 8.14
GMA PT 3 PNA 8.18
3752 a 3764 m   9°5'5.046" S 77°39'33.54" W
Observations: *Entoloma* Singer (1975:684): Not in Dennis (1970:76-80), nor Lazo (2001:146-153

P148 ***Entoloma sp.** (Clitopilopsis)?* Spores: x̄ (8.2x7.4)μm (Q=1.1)
Central Andes: Lima

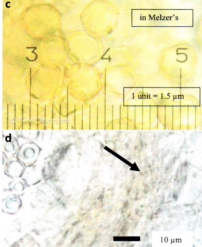

in Melzer's

1 unit = 1.5 μm

10 μm

Region: Zarate, Huarochirí, LIMA - Central Andes (figure e)
Ecology: organic material beside path to the Zarate *Polylepis* cloud forest (figure e)
Macroscopic
Pileus: 2.5cm, collybinoid - plane, slightly depressed, orange brown, smooth, with concentric zones and radial striate lines (figure a)
Stipe: (3.5-4)cm, light brown, smooth, centric, without annulus (fig. b)
Gills: cream, sinuate to sub-decurrent, distant-close (figure b)
Spore print: pink-orange
Microscopic
Spores (7.5-9.0)x(6.8-7.8)μm, x̄ (8.2x7.4)μm (Q=1.1) (n=5), 5-6 angled, thick walled, hyaline (KOH) non-amyloid (Melzer's), smooth, without germ-pore (figure c)
Trama:N.E.
Cystidia: not found
Basidia: lightly dextrose
Pileipellis: a trichoderm: of interwoven hyphae (white arrow) over what may be a subdermis of inflated cells (black arrow) - scalp section (figure d)
Sample: GMA PT 6 PCA 1.10
2928 m   11°55'55.079" S 76°28'53.448" W
Observations: *Entoloma*: but it does not fit well into the classic mold either in the key of Dennis (1970:76-80), nor in Lazo (2001:146-153). *Leptonia or Norlandia* have sub-decurrent gills), but the form does not correspond. Its like a *Leptonia* by pileus, habit, and size. The epicutis seems more *Nolanea*. Possibly *Clitopilopsis* with angular spores with wall more than 0.5μm

Entoloma sp. _(Clitopilus)?_   Esporas: x̄ (8.8x7.3)μm (Qm=1.2)
Central Andes: Lima

1 unit = 1.5 μm

1 unit = 6.5 μm

Region: Zarate, Huarochirí, Lima - Central Andes (figure g)
Ecología: in moss below shrubs (figure g)
Macroscopic
Pileus: (1-2.5)cm, convex- shallowly depressed, but otherwise shaped like _Entoloma_, cream, smooth, dry, with darker radial lines from top to margins (figures c,d)
Stipe: (6-7.5)cm, cream, smooth, slightly farinose close to gills, central, without annulus, and with white base (figure c)
Gills: cream, adnate to sub-decurrent, close (figure c)
Spore print: did not yield print
Microscopic
Spores: (7.5-9.0)x(6.8-7.5)μm, x̄ (8.8x7.3)μm (Qm=1.2) (n=5), irregular 5-6 angular transversal (red arrow) and longitudinal slightly angular (yellow arrow), smooth, without germ-pore (figure d)
Trama: parallel (blue arrow)
Cystidia: not found
Basidia: c.(20-24)x8, with 2-4 sterigmata (figure f: white arrow)
Pileipellis: a cutis: a carpet of undifferentiated repent hyphae (black arrow) (see Singer (1975) p.684.
Sample: GMA PT 6 PCA 1.4
2822m   11°55'48.726" S 76°29'17.682" W
Observations: _Entoloma:_ for reasons of its epicutis and color and size. But, it possibly _Rhodocybe_ (now _Clitopilus_) because spores comparatively only slightly angular. Not in Dennis (1970:75-80, Lazo (2001:146-153),

Entoloma sp.
Southern Andes: Cusco

Spores: x̄ (9.1x7.4)μm (Qm=1.2)

Region: Zurite, Anta, Cusco - Southern Andes (figure
Ecology: soil in *Polylepis besseri* cloud forest (figure e)
Macroscopic
Pileus: (1-2)cm, convex with level top, split, white, smooth, dry (figure a)
Stipe: 2-3(1)cm, white, fleshy, smooth, without annulus
Gills: white, sub-decurrent, close (figure b)
Spore print: did not yield print
Microscopic
Spores: (6.5-10.5)x(6.5-8) μm, x̄ (9.1x7.4)μm (Qm=1.2) (n=5), angular, hyaline (KOH), non-amyloid (Melzer's), without germ-pore (figure c)
Cystidia: not found
Pileipellis: a trichoderm?: of inflated hyphae that appear to originate vertically from subdermal cells - scalp section (figure d: black arrow)
Sample: GMA PT 3 PSA 4.19
3974 m   13°33'53.586" S 72°1'42.588" W
Observations: *Entoloma:* fleshy fungus and spores like *Entoloma*. The sub-current laminae are like *Eccillia* now is included within *Entoloma*.

1 unit = 3.75 μm

P151   ***Entoloma sp.***   Spores: x̄ (10.2x7.0)μm (Qm=1.4)
Southern Andes: Apurimac

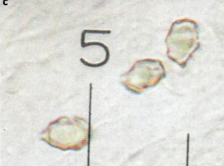

Region: Abancay, Apurimac - Southern Andes
Ecology: organic material in *Podocarpus glomeratus* cloud forest
Macroscopic
Odor: strong
Pileus: 1.5cm, conical, dark blue, shiny, atomate or squamous, moist (fig. a)
Stipe: (4-5)cm, dark blue, thin, smooth, central, without annulus (fig.b)
Gills: white-light blue, adnate-notched), close (figure b)
Spore print: not taken (one sample)
Microscopic
Spores: (8.5-11.5)x(5.5-7.5)μm, x̄ (10.2x7.0)μm (Qm=1.4) (n=5), (7-10) angular irregular, hyaline (KOH), non-amyloid (Melzer's), without germ-pore (figure c)
Cystidia: not found
Basidia: c.(37 x 10)μm
Epicutis: N.E.
Sample: GMA PT 3 PSA 2.19B
3224 m   13°35'50.004" S 72°52'42.006" W
Observation: *Entoloma sub Genus. Leptonia:* see also Dennis (1970:76) and *Rhodophyllus* (Singer 1975:672)

1 unit = 3.75 μm

P152 **_Entoloma sp._**   Spores: x̄ (9.7x8.0)µm (Qm=1.2)
Southern Andes: Cusco

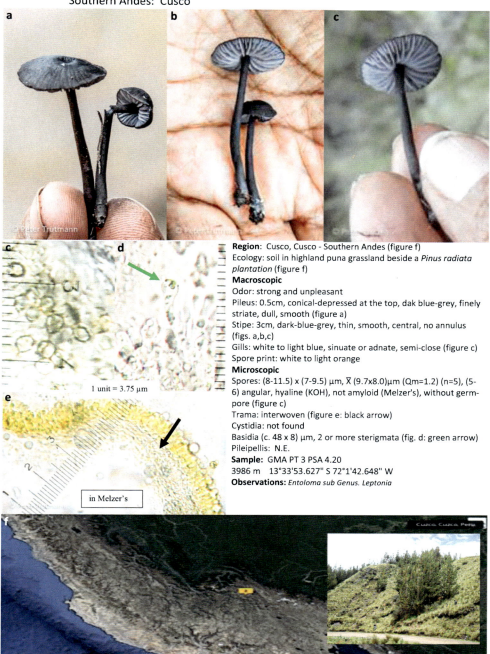

1 unit = 3.75 µm

in Melzer's

Region: Cusco, Cusco - Southern Andes (figure f)
Ecology: soil in highland puna grassland beside a _Pinus radiata plantation_ (figure f)
Macroscopic
Odor: strong and unpleasant
Pileus: 0.5cm, conical-depressed at the top, dak blue-grey, finely striate, dull, smooth (figure a)
Stipe: 3cm, dark-blue-grey, thin, smooth, central, no annulus (figs. a,b,c)
Gills: white to light blue, sinuate or adnate, semi-close (figure c)
Spore print: white to light orange
Microscopic
Spores: (8-11.5) x (7-9.5) µm, X̄ (9.7x8.0)µm (Qm=1.2) (n=5), (5-6) angular, hyaline (KOH), not amyloid (Melzer's), without germ-pore (figure c)
Trama: interwoven (figure e: black arrow)
Cystidia: not found
Basidia (c. 48 x 8) µm, 2 or more sterigmata (fig. d: green arrow)
Pileipellis: N.E.
Sample: GMA PT 3 PSA 4.20
3986 m   13°33'53.627" S 72°1'42.648" W
Observations: _Entoloma sub Genus. Leptonia_

THE MACROFUNGI OF ANDEAN PERU Part 1   179

Entoloma sp.   Spores: x̄ (10.2x8.6)μm (Qm=1.2)
Southern Andes: Cusco

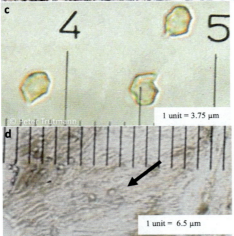

1 unit = 3.75 μm

1 unit = 6.5 μm

Region: Cusco, Cusco - Southern Andes (figure e)
Ecology: organic material in native Ichu puna grassland (fig. e)
Macroscopic
Odor: strong and unpleasant
Pileus: 2cm, plane-slightly depressed, dark blue-grey, velvet, dull (figure a)
Stipe: 5 cm, dark blue-grey, firm, smooth, central, no annulus (fig. b)
Gills: white-light blue, adnate, close (figure b)
Spore print: pink-orange (skin)
Microscopic
Spores: (8-11.5)x(7-7.5)μm, x̄ (10.2x8.6)μm (Qm=1.2) (n=5), (6-10) angular, hyaline (KOH), non-amyloid (Melzer's) without germ-pore (figure c)
Pileipellis: a cutis: made of a mycelial carpet -scalp section (figure d:black arrow)
Sample: GMA PT 3 PSA 4.32
4064 m   13°33'47.97" S 72°1'35.658" W
Observations: *Entoloma sub Genus. Leptonia*

Entoloma sp.   Spores: x̄ (10.5x 7.6)µm (Qm=1.4)
Southern Andes: Cusco

© Peter Trutmann

© Peter Trutmann

1 unit = 6.5µm

1 unit = 3.75 µm

Region: Ocra, Cusco - Southern Andes (figure f)
Ecology: in soil below native trees in degraded over used environment (figure f)
Macroscopic
Odor: pleasant, like potato
Pileus: 5cm, plane, dark blue-grey, smooth, shiny (figure a)
Stipe: 5cm, dark blue-grey, thin, smooth, central, without an annulus (figures a, b)
Gills: white - light blue, edges white, adnate - notched?, crowded (figure b)
Spore print: light orange (skin colored)
Microscopic
Spores: (8-11.5) x (7.5-8) µm, x̄ (10.5x 7.6)µm (Qm=1.4) (n=5), (5-10) angular, hyaline (KOH), non-amyloid (Melzer's), without germ-pore (figure c)
Trama: irregular-interwoven (figure e: white arrow)
Cystidia: not found
Basidia: with 2 or more sterigmata (figure e)
Pileipellis: a cutis: a carpet of brown pigmented mycelium - scalp section (figure d: black arrow)
Sample: GMA PT 3 PSA 5.15
3764 m   13°36'12.114" S 72°12'18.378" W
Observations: *Entoloma sub Genus Leptonia:* Traditional Quechua: yana kallampa

© Peter Trutmann

Google

Entoloma sp.

Southern Andes: Cusco

Spores: x̄ (11.2x7.6)µm (Qm=1.5)

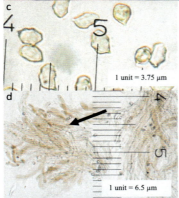

1 unit = 3.75 µm

1 unit = 6.5 µm

Region: Zurite, Anta, Cusco - Southern Andes (figure e)
Ecology: in soil below _Polylepis besseri_ cloud forest (figure e)
Macroscopic
Odor: mild
Pileus: (2.5-4)cm, plane-broadly convex, dark blue-grey, smooth, shiny, dry (figure a)
Stipe: 5-7(0.4)cm, dark blue-grey, lighter at lower end, smooth, central, without annulus (figure b)
Gills: white, with fine blue edge, sinuate-adnate, close to crowded (figure c)
Spore print: very light brown - cream
Microscopic
Spores: (8-12)x(7-8.5)µm, x̄ (11.2x7.6)µm (Qm=1.5) (n=5), (4-10) angular, hyaline (KOH) non-amyloid (Melzer's) without germ-pore (figure)
Cystidia: N.E.
Basidia: N.E.
Pileipellis: a demis - a trichoderm? of what appear to be relatively specialized pigmented hyphae - scalp section (figure)
Sample: GMA PT 3 PSA 7.16,
           GMA PT 3 PSA 7.20CF.
4012 m   13°25'48.197" S 72°15'24.767" W
Observations: _Entoloma sub Genus Leptonia_

Entoloma sp.
Southern Andes: Cusco

Spores: x̄ (9.1x6.2)μm (Qm=1.5)

Region: Zurite, Anta, Cusco - Southern Andes (figure g)
Ecology: in soil below *Polylepis besseri* cloud forest (figure g)
Macroscopic
Odor: mild
Pileus: (2-3)cm azu, broadly convex mildly depressed, dark blue-grey, dry, shiny, very lightly, finely wrinkled (figures a,b)
Stipe: 5-7(0.2-0.4)cm, light to dark-blue brown or grey, smooth, central, without annulus (figures b,c)
Gills: white-pink, adnexed-notched?, close
Spore print: did not yield print
Microscopic
Spores: (7.5-11.5) x (5.5-7.5) μm, x̄ (9.1x6.2)μm (Qm=1.5) (n=5), (5-8) angular, hyaline (KOH), non-amyloid (Melzer's), without germ-pore (figure d)
Cystidia: not found
Basidia: c.(30 x 8 μm) with 2 or more sterigmata (figure e: white arrow)
Pileipellis: a cutis: a carpet of undifferentiated repent hyphae - scalp section (figure f: black arrow)
Sample: GMA PT 3 PSA 7.23
4049 m   13°25'46.53" S 72°15'24.989" W
Observations: *Entoloma sub Genus Leptonia*

1 unit = 3.75 μm

1 unit = 33 μm

Entoloma sp.
Central Andes: Junín

Spores: x̄ (9.1x7.4)μm (Qm=1.2)

Region: Huancayo, Junín - Central Andes (figure g)
Ecology: soil in highland Ichu puna grassland (figure g)
Macroscopic
Pileus: 1cm, umbonate convex, black, smooth, dry, shiny (figures a,b)
Stipe: 1.5cm, white-pink, with white streaks, darker and slightly farinose below gills (spores?), smooth, without annulus (figures d,c)
Gills: pink, sinuate, eroded, close - sub distant (figure c)
Spore print: did not yield print
Microscopic
Spores: (8.2-11.2)x(7.0-7.9)μm, x̄ (9.1x7.4)μm (Qm=1.2) (n=5), irregularly 5-7 angled, hyaline (KOH) inamyloid to slightly amyloid (Melzer's), without germ-pore (figures, d, e,f)
Trama: parallel (figure e: white arrow)
Cystidia: not found
Basidia: c.(26x6)μm with two or more spores, slightly dextrose? (fig.f: black arrow)
Pileipellis: a cutis: mycelium
Sample: GMA PT 3 PCA 1.13
4298 m   11°58'22.823" S 75°3'59.327" W
Observations: *Entoloma* sub Genus *Leptonia*

1 unit = 3.75 μm

in Melzer's

1 unit = 3.75 μm

Entoloma sp.
Central Andes: Ancash

Spores: x̄ (10.7x7.5)µm (Qm=1.4)

Region: Ilanganuco, Yungay, Ancash - Central Andes (figure g)
Ecology: soil below *Polylepis besseri* cloud forest (figure g)
Macroscopic
Pileus: 0.75cm, convex, dark blue, atomate, dry, with light margin (fig.a)
Stipe: 2cm, dark blue, smooth, with farinose base, central, without annulus (figures a,b)
Gills: white, thick, sinuate or emarginate, distant-close (figure b)
Spores: did not yield spores
Microscopic
Spores: (9.5-11.2)x(7.5)µm, x̄ (10.7x7.5)µm (Qm=1.4) (n=5), irregular (5-7) angles, hyaline (KOH) inamyloid (Melzer's), without germ-pore (fig. c)
Cystidia: not found
Basidia: c.(26x7.5)µm, inamyloid, with 4 spores (figure d: white arrow)
Pileipellis: a cutis of parallel partially pigmented repent hyphae - scalp section (figure f:black arrow)
Sample: GMA PT 5 PNA 1.2.
est. 3700m   9°5'25.476" S 77°40'41.07" W
Observations: *Entoloma sub Genus Leptonia*

P159

Entoloma sp.
Central Andes: Ancash

Spores: x̄ (10.9x7.8)µm (Qm=1.4)

Region: Llanganuco, Yungay, Ancash - Central Andes (figure e)
Ecology: organic matter, by track in *Polylepis besseri* cloud forest
Macroscopic
Pileus: 5cm, mammilate, cream and central calotte brown, finely corrugated, with broadly undulating margins (figure a)
Stipe: 9cm, cream, smooth, central, without annulus (figures a,b)
Gills: cream, adnexed, close to crowded (figure b)
Spore print: did not yield print
Microscopic
Spores: (9 5-11.2)x(7.1-9.4)µm, x̄ (10.9x7.8)µm (Qm=1.4) (n=5), prismatic, irregular (5-8) angles, hyaline (KOH), inamyloid (Melzer's), without germ-pore (figure c)
Pileipellis: A dermis - with setae or pileocystidia (black arrow) with a hyphal layer above a large celled subdermis? (white arrow) - scalp section (figure d)
Sample: GMA PT 4 PNA 1.4
3840m   9°4'44.705" S 77°39'9.395" W
Observation: *Entoloma subGenus Nolanea*: due to pileus form, and pileipellis with setae

1 unit = 3.75 µm

Entoloma sp.
Spores: x̄ (11.0x7.3)μm (Qm=1.5)

Northern Andes: Lambayeque

Region: Cañaris, Ferreñafe, Lambayeque - Northern Andes (figure h)
Ecology: wood, on a living trunk in relic Amazonian highland cloud forest of Cañaris (figure h)
Macroscopic
Odor: acidic unpleasant
Pileus: (1.5-4)cm, plane to mammilate, brown with darker calotte, smooth, moist to viscid (figure a,b)
Stipe: 4cm, brown, smooth, centric, without annulus (figures b,c)
Gills: light pink, emarginate or adnexed, crowded (figure c)
Spore print: did not yield print
Microscopic
Spores: (8.25x13.1)x(5.6-7.9)μm, x̄ (11.0x7.3)μm (Qm=1.5) (n=5), irregular (5-8)angular, hyaline (KOH), inamyloid (Melzer's) without germ-pore (figure d)
Trama: irregular (figure f: white arrow)
Cystidia: not found
Basidia: c. (48x15)μm with 2-4 spores, vacuoles (fig. e:yellow arrow)
Pileipellis: a cutis: of enlarged parallel hyphae - scalp section (figure g)
Sample: GMA PT 3 PNA 2.13
2721 m   6°3'42.755" S 79°15'7.206" W
Observations: *Entoloma subGenus Nolanea*: see Singer (1975:684). Not found in Dennis (1970:76-80). like a *Nolanea* by its fragile form, color, and convex than conical habit although at times umbonate like a *Leptonia* (which is almost never brown).

1 unit = 3.75 μm

1 unit = 6.5 μm

P161   *Entoloma sp*   Spores: x̄ (13.5x10,2)µm (Q=1.3)
Northern Andes: Lambayeque

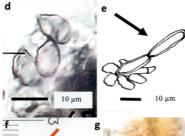

Region: Cañaris, Ferreñafe, Lambayeque - Northern Andes (figure h)
Ecology: organic material below shrubs (figure h)
Macroscopic
Pileus: 1cm, narrowly convex, orange red, micalaceous, like innate fibrils (figure a,b)
Stipe: (3-4)cm, orange red, with fine fibrous scales, central, with base of white mycelium containing rhizome-like strands (white arrow), without annulus (figures a,b,c)
Gills: white, emarginate or sinuate, distant (figure c)
Spore print: did not yield print
Microscopic
Spores: (11.6-15.0)x(7.9-11.3)µm x̄ (13.5x10,2)µm (Qm=1.3) (n=6), irregular 5-7 angular, hyaline (KOH), inamyloid (Melzer's) without germ-pore (figure d)
Trama: parallel? (figure g: yellow arrow))
Cystidia: none found
Basidia: c. (40x8) µm, with 2 or more spores /sterigmata (figure f)
Pileipellis: a dermis: like a cystoderma with enlarged c. (50x15)µm sub pellis with Pileocystidia (figure e: black arrow)
Sample: GMA PT 6 PNA 2.4
1450 m   6°1'54.66" S 79°12'8.322" W
Observation: *Entoloma subGenus Pouzarella*: Like (Baroni et al., 2012) from northern Argentina (13-19.4 x 8.5-11.3)µm (n= 40, 17.2 ± 1.33 x 10.4 ± 0.67, Q = 1.31-1.92, Qm = 1.65 ± 0.10). Caulocystidia not evaluated

Family Hydnangiaceae

Laccaria

Laccaria *cf. amethystina*   Spores: none found
Southern Andes: Cusco

Region: Zurite, Anta, Cusco- Southern Andes (figure d)
Ecology: organic material and moss under *Polylepis besseri* cloud forest (figure d)
Macroscopic
Pileus: (5-7)cm, uplifted, cream to violet, fragile, almost translucent, gelatinous, moist (figure a)
Stipe: (5-6)cm, cream, viscous, smooth, central, without annulus (figure a)
Gills: light purple, thick, adnate, close (a)
Spore print: did not yield print
Microscopic
Spores: not found
Trama: regular (figure c: white arrow)
Cystidia: not found
Basidia: c. (24-28)x(4-8)μm (figure b: red arrow)
Pileipellis: not evaluated
Samples: GMA PT 2 PSA 6.4
3916 m   13°25'55.139" S 72°15'18.6" W
Observations: tentatively placed in *Laccaria*: has likeness to L. *amethystin: Arora 1980:172, Reported from Costa Rica (Mata et al., 2003)*

1 unit = 3.75 μm

P165   ***Laccaria*** *(species complex?)*   Spores: x̄ (7.0x7.9) (Qm=1.0)

Southern Andes: Ancash, Apurimac, Cajamarca, Cusco, Lima and Puno

Region: Ancash, Apurimac, Cajamarca, Cusco, Lima, Puno - General Andes (fig. i)
Ecology: soil associated with Eucalyptus trees (figures a,i)
Macroscopic
Odor: agreeable
Pileus: (1-3)cm , variable -depressed plane or convex to uplifted when older, orange to sometimes pinkish, dry, narrowly striate (figures, a,b, c, d)
Stipe: (2-4)cm, orange with lighter color at base sometimes with mycelium, +/-fibrous smooth, central, without annulus (figure c)
Gills: white to lightly colored, thick, rough edged, adnexed to emarginate, distant to close (figure d)
Spore print: white (figure c: orange arrow)
Microscopic
Spores: (7.0-11.0)x(7.0-10.6)μm, x̄ (8.0x7.9) (Qm=1.0) (n=18), globose, echinulate, hyaline (KOH), inamyloid (Melzer's), without germ-pore (figure e)
Trama: parallel (figure g: white arrow)
Cystidia: not found
Basidia: c. (22-25x8)μm, predominantly with 2 sterigmata (some maybe 4 spored) (figure f: red arrow)
Pileipellis: a cutis: of pigmented inflated hyphae - scalp section (figure h: black arrow)

Samples: Huancane, Puno: GMA PT 1 PSA 10.5
3883 m   15"10'34.595" S 69"50'41.532" W
Putina, Puno: GMA PT 3 PSA 8B.3F
3977 m   14"54'40.219" S 69"52'28.079" W
Canchis, Cusco   GMA PT 1 PSA 3.2, GMA PT 1 PSA 3.3
3413 m   13"58'21.521" S 71"29'36.437" W
Abancay, Apurimac: GMA PT 3 PSA 2.11
3153 m   13"35'53.687" S 72"52'41.921" W
Canta, Lima: GMA PT 1 PCA 0.8F
3541 m   11"27'46.122" S 76"35'34.398" W
Yungay, Ancash GMA PT 3 PNA 7.4
GMA PT 3 PNA 7.5 GMA PT 3 PNA 7.6
3030 m   9"28'2.009" S 77"32'4.02" W
Cajamarca,Cajamarca: GMA PT 3 PNA 5.15
2706 m   7"13'58.667" S 78"16'6.179" W
Cumbemayo, Cajamarca GMA PT 3 PNA 3.2
3185 m   7"13'49.872" S 78"29'14.255" W

Observations: *Laccaria* (Singer 1975:230-231, Arora 1986:171). It appears to be a complex predominantly of a two spored species, perhaps *L. fraterna* (Pegler, 1965) possibly introduced from Australia with the *Eucalyptus* species. A four spored species found in Cajamarca may be *L. laccata* as the spore size was (7.1-11)x(7.1-.10.5)μm like L. laccata, compared to (7-9)x(7-8.9) μm further South.

1 unit = 3.75 μm   1 unit = 1 μm   1 unit = 3.75 μm

Laccaria** cf. **proxima
Northern Andes: Cajamarca

Spores: x̄ (8.1x7.4) µm (Qm=1.1)

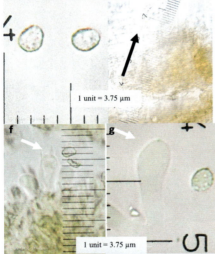

1 unit = 3.75 µm

1 unit = 3.75 µm

Region: Cumbemayo, Cajamarca - Northern Andes (figure h)
Ecology: soil below *Pinus radiata* (figure h)
Macroscopic
Odor: pepper
Pileus: (2-3)cm, plane depressed, orange, smooth, firm, without fine
corrugation (figure a,b)
Stipe: 5-6 cm, orange lighter at base, with light stripes, central,
without annulus (figures b,c)
Gills: white, thick, roughened edges, emarginate to adnate, close (fig.c)
Spore print: did not yield spore print
Microscopic
Spores: (7.5-9.4)x(6.8-7.5)µm, x̄ (8.1x7.4)µm (Qm=1.1) (n=10), oval,
echinulate, hyaline (KOH), inamyloid (Melzer's), no germ-pore (fig. d)
Cystidia: no evaluated
Basidia: c.(22-24)x(7-8)µm (fig.g), 2-3 sterigmata (fig. f:white arrow)
Pileipellis: a cutis: of parallel hyphae with pigmented suprapellis? -
scalp section (figure e: black arrow)
Samples:
Cumbemayo, Cajamaraca: GMA PT 3 PNA 4.2
3475 m 7°11'42.197" S 78°33'12.923" W
Cajamarca, Cajamarca, GMA PT 3 PNA 4.4 3650 m 7°11'18.791" S
78°34'34.518" W
Observations: *Laccaria*: Singer (1975:230-231) Arora (1986:172). A
species with 2-3 sterigmata, oval spores to subglobose smaller than
those under *Eucalyptus*. *Not L.laccata* which has 4 spored basidia.

THE MACROFUNGI OF ANDEAN PERU   Part 1

Family Hygrophoraceae

(Lodge et al.
Hyg...yboideae, Tri...

© Peter Trutmann

Chromosera

Chromosera sp.
Central Andes: Ancash

Spores: x̄ (9.7x5.3)μm (Qm=1.8)

Region: Yungay, Ancash - Central Andes (figure g)
Ecology: organic matter and moss in *Polylepis besseri* dominated cloud forest (figure f)
Macroscopic
Pileus: (1-2)cm, narrowly convex and depressed, yellow, smooth, mildly corrugated, dry (figure a)
Stipe: (2-3)cm, central, yello, smooth, without annulus (figures a,b)
Gills: yellow, wide, decurrent, distant
Spore print: did not yield print
Microscopic
Spores: (7.9-11.3)x(4.5-6.0)μm, x̄ (9.7x5.3)μm (Q=1.8) (n=5), elliptical-amygdaliform, hyaline (KOH), smooth, inamyloid (Melzer's) without germ-pore (figure c)
Trama: irregular (divergent?), pseudoamyloid (figure d: black arrow)
Cystidia: not found
Basidia: c. (37.5x4)μm, 4 sterigmata, slightly pseudoamyloid (fig. d)
Pileipellis: a cutis: a suprapellis carpet of hyphae above a subpellis of pigmented repent hyphae (white arrow) above hyaline hyphae (black arrow) - scalp section (figure e)
Sample: GMA PT 4 PNA 1.9
3864m   9°4'41.255" S 77°39'9.539" W
Observations: conforms to description of *Chromosera*.

in Melzer's

1 unit = 3.75 μm

Gliophorus

P168F   *cf. **Gliophorus sp** .*   *Photo record only*
Northern Andes: Lambayeque

Región: Cañaris, Ferreñafe, Lambayeque - Northern Andes (figure d)
Ecology: organic material in relic Amazonian highland cloud forest (figure d)
Macroscopic
Pileus:(1-2)cm, plane- umbonate, orange, smooth, viscid (figures a,b)
Stipe: (5-6)cm, yellow, fibrous, smooth, viscid, no annulus or volva (fig. c)
Gills: white-light yellow, adnate-sub-decurrent?, close (figure c)
Spore print: not taken
Photo record: GMA PT 5 PNA 2.25F
2415m   6°4'24" S 79°14'3" W
Observations: Tentatively identified as a *Gliophorus*: by form, viscocity.and slimy gill edges. However, also keys out to *H.parvilus* Peck (Dennis 1970) p.18. found in El Avila, Venezuela and similar to *H. noninquinans* reported from Mato Grosso en Brazil (Sourell et al., 2018), and in Peru by Pavlich (1976) and Alvarez et al (2014).

Gliophorus sp.
Central Andes: Ancash

Spores: (x̄ (11.1x7.3)μm (Qm=1.5)

Region: Recuay, Ancash - Central Andes (figure f)
Ecology: soil in highland Ichu puna grasslands (figure f)
Macroscopic
Pileus: (1-1.5)cm, broadly convex, orange-red, smooth, moist (figure a,b)
Stipe: (2-3)cm, yellow, central, smooth, moist without annulus (figure b)
Gills: yellow, emarginate, subdistant (figure c)
Spore print: did not yield print
Microscopic
Spores: (10.5-11.5)x(6.7-7.5) μm x̄ (11.1x7.3) μm (Qm=1.5) (n=2), few, elliptical, smooth, hyaline (KOH), lightly amyloid (Melzer's) (figure d:arrow)
Trama: N.E.
Cystidia: N.E.
Basidia: c.(40x8)μm, elongated, inamyloid, releasing oil globules in Melzer's (figure e: arrow)
Pileipellis: not evaluated
Sample: GMA PT 3 PNA 9.4
3872 m   9°56'34.986" S 77°22'55.626" W
Observations: *Gliophorus:* by its viscid cap and slightly amyloid staining spores

in Melzer's

5 μm

1 unit = 3.75 μm

THE MACROFUNGI OF ANDEAN PERU   Part 1

cf. **Gliophorus sp.**   Photographic record only
Central Andes: Ancash

Region: Llanganuco, Yungay, Ancash - Central Andes (figure d)
Ecology: organic material in *Polylepis besseri* dominated cloud forest (figure d)
Macroscopic
Pileus: (1-2)cm, conical, red-orange, smooth, older specimens with pale or dark scale like covering, viscid (figures a, b, c)
Stipe: (3-6)cm, thick compared to cap, reddish orange, central, moist to viscid, without annulus (figure a,b,c)
Gills: white - light orange, waxy, adnate emarginate, distant
Spore print: not taken
Sample: GMA PT 5 PNA 1.6DF
3847 m   9°4'43.134" S 77°39'8.292" W
Observation: Tentatively a *Gliophorus:* by its viscosity and *Hygrocybe* like shape.

Humidicutis

Humidicutis sp.
Northern Andes: Lambayeque

Spores: x̄ (7.8x5.1)µm (Qm=1.5)

© Peter Trutmann

1 unit = 1.5µm

10 µm

1 unit = 3.75µm

Region: Cañaris, Ferreñafe, Lambayeque -Northern Andes (figure i)
Ecology: organic material in relic northern Amazonian cloud forest (figure i)
Macroscopic
Pileus: 4cm, umbonate, separated into sections with age, white, smooth, dry to lightly humid (figure a)
Stipe: (10-12)cm, white, hollow, smooth, without annulus or volva (figures a,b,c)
Gills: white, serrated, emarginate-sinuate or sub-decurrent, close (figure c)
Spore print: did not yield print
Microscopic
Spores: (7.5-9.0)x(5.0-5.3)µm, x̄ (7.8x5.1)µm (Q=1.5) n=(5), elliptical, smooth to lightly ornamented, hyaline (KOH), inamyloid (Melzer's), no germ-pore (figure d)
Trama: parallel -without 'clamp' connections (figure g)
Cystidia: not found
Basidia: c.(26-33)x(4-6)µm, elongated with 2 or perhaps more sterigmata (figures e,f: white arrow)
Pileipellis: a cutis or trichoderm (uncertain): of inflated parallel hyphae (black arrow) in a gelatinous medium -scalp section (figure h)
Sample: GMA PT 6 PNA 3.6
2707m   6°3'42.221" S 79°15'8.142" W
Observations: *Humidicutis*: due to carpophore form and white pigment, gills, spore size. It is not *H. subaustralis* found in Trinidad (Dennis 1970:16.) Has similarity to a *Hygrocybe* sp. 1 Mato Grosso, Brasil (Sourell et al 2018: imagines 240-1). Not found in literature from Peru, Colombia or Ecuador.

© Peter Trutmann

THE MACROFUNGI OF ANDEAN PERU Part 1

Neohygrophorus

P172   ***Neohygrophorus** sp.*   Spores: x̄ (5.4x4.5)µm (Qm=1.2)
(3)   Southern Andes: Arequipa and Puno

Region: Lampa, Puno and Caylloma, Arequipa - Southern Andes (figure j)
Ecology: soil, in highland puna grasslands (figures h,i)
Macroscopic
Odor: agreeable
Pileus: (3-6)cm, broadly convex to plain, light brown to brown, smooth, corrugated around margins, moist, dull (figures a,b)
Stipe: 5-10(0.5)cm, white, smooth, fibrous, without annulus (figure c)
Gills: white, thick, waxy, emarginate-adnexed, distant (figure c)
Spore print: did not yield print
Microscopic
Spores: (4.1-6.8)x(3.8-5.0)µm, x̄ (5.4x4.5)µm (Qm=1.2) (n=11), subglobose, with thin wall, smooth, hyaline (KOH), lightly amyloid (Melzer's), without germ-pore (figure d)
Trama: parallel, broad, with clamp connections, reddening in KOH (figure f: white arrow)
Cystidia: not found
Basidia: elongated with two or more spores (figure e: red arrow)
Epicutis: a dermis?: in a gelatinous matrix (figure g) with epiphytes (fig with: blue arrow)
Sample: Lampa, Puno GMA PT 1 PSA 0.4
4398 m   15°43'51.503" S 70°50'57.215" W
Caylloma, Arequipa: GMA PT 2 PSA 10.8
4454 m   15°51'39.773" S 71°7'33.804" W
Observations: *Neohygophorus*: (see Singer 1975:206 and Laessoe and Petersen 2019: 143) due to amyloid spore reaction and parallel trama and *Hygrocybe* waxy gills characteristics.

d: in Melzer's
e
1 unit = 3.75µm
f
g: in Melzer's

Neohygrophorus sp.   Spores: x̄ (7.3x4.2)μm (Qm=1.7)
Southern Andes: Puno

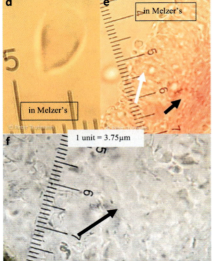

in Melzer's

in Melzer's

1 unit = 3.75μm

Region: Huancane, Puno - Southen Andes (figure g)
Ecology: soil in rocky puna with Ichu on a barren mountain (figure g)
Macroscopic
Odor: N.E.
Pileus: (1-4)cm, broadly convex to plane, cream darkening slightly at
the center, smooth, moist, dull, with eroded margins (figures a,b)
Stipe: (3-4)cm, white, smooth, central, without annulus (figure c)
Gills: white, thick, waxy, emarginate, subclose (figure c)
Spore print: did not yield a print
Microscopic
Spores: (6.0-9.0)x(4.0-5.0)μm, x̄ (7.3x4.2)μm (Qm=1.7) (n=6),
elliptical, thin walled, smooth, hyaline (KOH), lightly amyloid
(Melzer's) without germ-pore (figure d)
Trama: parallel, red in (Melzer's) (figure e: black arrow)
Cystidia: not found
Basidia: c.(30 x 6)μm, with 2 spores (flecha blanca)
Pileipellis: a dermis: of what appear as short, broad, hyphal tips -
scalp section (figure f: black arrow)
Sample: GMA PT 1 PSA 10.9
4062 m   15°10'10.824" S 69°50'58.457" W
Observations: *Neohygrophorus*: (see Singer 1975:206) due to
amyloid spore reaction with otherwise many *Hygrophorus*
characters. Removal from Hygrophoraceae was proposed by
Matheny et al (Matheny et al., 2006). Syn. with *Pseudoomphalina*-
Tricholomataceae (Vizzini and Ge, 2015)

Hygrocybe

P174   ***Hygrocybe sp.*** sect. Hygrocybe   Spores: x̄ (11.4x7.4)µm (Qm=1.5)
Southern Andes: Cusco

Region: Zurite, Anta, Cusco - Southern Andes (figure g)
Ecology: soil and organic material in Polylepis besseri cloud forest (fig. h)
Macroscopic
Odor: mild and agreeable
Pileus: (3-7.5)cm, broadly conical with characteristic apical pointed umbo, red turning black closer to center, smooth, dry (figure a,b)
Stipe: (6.5-12)cm, orange red when young then more yellow orange red , streaked, smooth, dry, central, without annulus (figures a,b,c)
Gills: white-light yellow, slightly serrated and waxy, adnate to lightly sub-decurrent, close (figure c)
Spore print: white
Microscopic
Spores: (11.0-11.5)µm x (6.8-8)µm, x̄ (11.4x7.5)µm (Qm=1.5) (n=5), elliptical, smooth, with inclusions or slightly ornamented (KOH), hyaline (KOH) inamyloid (Melzer's) without germ-pore (figure d)
Trama: parallel (figure f: black arrow)
Cystidia: not found
Basidia: c.(28-32)x(4)µm, with at least two sterigmata (figure e)
Pileipellis: a cutis: a carpet of mycelium in a gelatinous medium
Sample: GMA PT 3 PSA 7.17,
         GMA PT 3 PSA 7.28F
4031 m   13°25'48.024" S 72°15'24.654" W
         GMA PT 3 PSA 7.26
4057 m   13°25'46.145" S 72°15'25.211" W
Observations: *Hygrocybe*: characterized from others by its large size, peaked cap and red-black coloration, spore size and basidium size.

1 unit = 3.75µm

1 unit = 6.5µm

Hygrocybe sp. sect. _Hygrocybe_   Photographic record only
Southern Andes: Cusco

Region: Zurite, Anta, Cusco - Southern Andes (figure c)
Ecology: organic material in _Polylepis besseri_ cloud forest
(figure c)
Macroscopic
Pileus: 5cm, conical, orange with pronounced apical umbo,
smooth, moist, and shiny (figures a, b)
Stipe: 14(1)cm, broad, white-yellow turning black with age,
smooth, without annulus (figure b)
Gills: white, uncertain adhesion to pileus or density
Photo sample:
GMA PT 3 PSA 7.26BF UNSAAC
4991 m   13°25'50.142" S 72°15'22.632" W
Observations: _Hygrocybe_: characterized by an accentuated
apical umbo and large size as well as having an unextended
cap at maturity. It is likely a variation of GMA PT 3 PSA 7.17
from same location. However, it differs by color and lack of
an extended broadly conical cap at maturity.

Hygrocybe sp. sect. Hygrocybe   Spores: x̄ (11.3x7.5)µm (Qm=1.5)
Southern Andes: Cusco

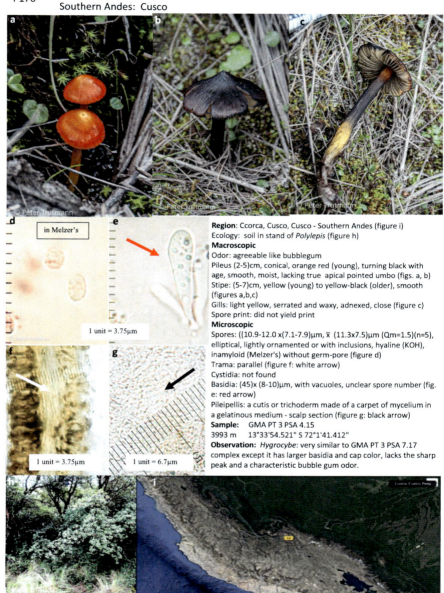

Region: Ccorca, Cusco, Cusco - Southern Andes (figure i)
Ecology: soil in stand of *Polylepis* (figure h)
Macroscopic
Odor: agreeable like bubblegum
Pileus (2-5)cm, conical, orange red (young), turning black with age, smooth, moist, lacking true apical pointed umbo (figs. a, b)
Stipe: (5-7)cm, yellow (young) to yellow-black (older), smooth (figures a,b,c)
Gills: light yellow, serrated and waxy, adnexed, close (figure c)
Spore print: did not yield print
Microscopic
Spores: ((10.9-12.0 x(7.1-7.9)µm, x̄ (11.3x7.5)µm (Qm=1.5)(n=5), elliptical, lightly ornamented or with inclusions, hyaline (KOH), inamyloid (Melzer's) without germ-pore (figure d)
Trama: parallel (figure f: white arrow)
Cystidia: not found
Basidia: (45)x (8-10)µm, with vacuoles, unclear spore number (fig. e: red arrow)
Pileipellis: a cutis or trichoderm made of a carpet of mycelium in a gelatinous medium - scalp section (figure g: black arrow)
Sample: GMA PT 3 PSA 4.15
3993 m   13°33'54.521" S 72°1'41.412"
Observation: *Hygrocybe:* very similar to GMA PT 3 PSA 7.17 complex except it has larger basidia and cap color, lacks the sharp peak and a characteristic bubble gum odor.

in Melzer's

1 unit = 3.75µm

1 unit = 3.75µm

1 unit = 6.7µm

Hygrocybe sp. sect. Hygrocybe   Spores: x̄ (11.2x7.7)µm (Qm=1.4)
Southern Andes: Cusco

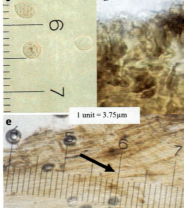

1 unit = 3.75µm

Region: Anta, Cusco - Southern Andes (figure f)
Ecology: soil below native stand of trees in degraded environment (figure f)
Macroscopic
Odor: mild and agreeable
Pileus: (2-3)cm, conical with accentuated apical pointed umbo, orange- brown darkening to center, smooth, moist, eroded margins , (fig. a)
Stipe: 10cm, yellow-white with greyish lesions with age, smooth, without annulus (figure b)
Gills: white, serrated and waxy, emarginate?, close (figure b)
Spore print: white
Microscopic
Spores: (10.5-11.5)x(7-8)µm, x̄ (11.2x7.7)µm (Qm=1.4) (n=5), subglobose, with inclusions, smooth, hyaline (KOH) inamyloid, without germ-pore (figure c)
Trama: parallel
Cystidia: not found
Basidia: with dark inclusions, 2-4 sterigmata (fig. d: white arrow)
Pileipellis: a cutis: parallel hyphae in gelatinous medium - scalp section (figure
Sample: GMA PT 3 PSA 5.7
3745 m   13°36'11.861" S 72°12'20.495" W
Observations: *Hygrocybe*: consumed locally 'Huano chuchuca'

THE MACROFUNGI OF ANDEAN PERU Part 1

Hygrocybe sect. Hygrocybe Spores: x̄ (12.1x6.0)µm (Qm=2.0)
Central Andes: Cerro De Pasco

c 1 unit = 1.5µm

d 1 unit = 3.75µm

e 1 unit = 6.5µm

f 1 unit = 1.5µm

Region: Vicco, Cerro de Pasco - Central Andes (figure g)
Ecology: soil in highland puna Ichu grasslands (figure g)
Macroscopic
Odor: N.E.
Pileus:(2 -2.5)cm, conical with an apical pointed umbo, orange red darkening black towards center, smooth, slightly moist (figure a)
Stipe: (3-5)cm, yellow to more orange maturing and darkening with age. smooth, hollow, central, without annulus (fig.b)
Gills: white- light yellow, waxy, emarginate,-adnate even free, close to subclose (figure b)
Spore print: did not yield print
Microscopic
Spores: (11.3-13.5)x(5.3-6.3)µm, x̄ (12.1x6.0)µm (Qm=2.0) n=(5), oblong, light brown (KOH), inamyloid (Melzer's), smooth, without germ-pore (figure c)
Trama: made of short, inflated cells like bricks that appear parallel (figure e: white arrow)
Cystidia: not found
Basidia: c.(30-38 x4)µm, elongated and thin, with 4 spores (figure f; red arrow).
Pileipellis: a cutis: of pigmented hyphae - radial section (figure d: black arrow)
Sample: GMA PT 7 PCA 2.10
4151 m   10°50'26.94" S 76°10'37.295" W
Observations: *Hygrocybe*: keys to *H.conica* en Arora (1989:116).
Characteristics of the sample are the lightly brown, elongated spores

Hygrocybe sect. Hygrocybe   Spores: ($\bar{x}$ (10.6x6.4)μm (Qm=1.6)
Central Andes: Cerro De Pasco

Region: La Quinua, Cerro de Pasco (figure h)
Ecology: soil in grassland associated with *Polylepis besseri* cloud forest (figure h)
Macroscopic
Odor: agreeable strong (dry)
Pileus: 3cm, broadly conical, with an apical pointed umbo , reddish orange with a black central peak, smooth slightly striate (figures a,b)
Stipe: 4cm, yellow, with orange discoloration with age, smooth, slightly moist, central, without annulus (figure b)
Gills: light yellow-orange, waxy, emarginate-adnate, close (fig. c)
Spore print: did not yield print
Microscopic
Spores: (10.5-12.0)x(6.0-7.5)μm, $\bar{x}$ (10.6x6.4)μm (Qm=1.6) (n=5), elliptical to oblong, smooth, hyaline (KOH), inamyloid (Melzer's) without germ-pore (figure d)
Trama: parallel
Cystidia: cheilocystidia? (figure f:white arrow)
Basidia: c.(33 x8)μm, elongated and thin, with 4 spores (fig. g: red arrow)
Pileipellis: a trichoderm: of perpendicular pigmented hyphae - radial section (figure e: black arrow)
Sample GMA PT 7 PCA 3.1
3733m   10°38'18.413" S 76°10'14.639" W
Observations: *Hygrocybe:* keys out to the omnipresent *H.conica* see Arora (1989:116). However, the sample appears be characterized by an orange cap with little central black, with cystidia sometimes present in *Hygrocybe* (Singer 1975:207)

1 unit = 1.5μm

1 unit = 3.75μm

1 unit = 3.75μm

Hygrocybe sp. sect. Hygrocybe   Spores: x̄ (13.3x7.2)µm (Qm=1.8)
Southern Andes: Cusco

Region: Chinchero, Cusco - Southern Andes (figure f)
Ecology: soil in pasture with *Eucalyptus* (figure e)
Macroscopic
Odor: mild (dry)
Pileus (3-4)cm, broadly conical to upturned umbonate, light
pink-red with a black central calotte, smooth, dry (figure a)
Stipe: 4-6(0.5)cm, pink to red, white closer to base, dry, smooth,
central without annulus (figure a)
Gills: yellow-orange, waxy, slightly serrated, adnexed to lightly
decurrent, close (figure a)
Spore print: white
Microscopic
Spores: (12.0-15.0)x(6.7-7.5)µm, x̄ (13.3x7.2)µm (Qm=1.8) (n=5),
elliptical to oblong, hyaline (KOH), inamyloid (Melzer's), smooth,
without germ-pore (figure b)
Trama: parallel (figure d: white arrow)
Cystidia: none found
Basidia: c. (32)x (4-7)µm, with 2 or more spores (figure c: black
arrow)
Pileipellis: a cutis: of repent hyphae in gelatinous substrate
Sample: GMA PT 3 PSA 3.20
3792 m   13°23'20.874" S 72°2'24.299" W
Observations: *Hygrocybe*: sample characterized by larger spores
and cap color (which may have been washed out).

1 unit = 3.75µm

1 unit = 33µm.

Hygrocybe sp. sect. Hygrocybe   Spores: x̄ (11.5x7.6)µm (Qm=1.5)
Central Andes: Cerro De Pasco

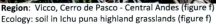

Region: Vicco, Cerro de Pasco - Central Andes (figure f)
Ecology: soil in Ichu puna highland grasslands (figure f)
Macroscopic
Pileus: (1-4)cm, broadly conical, lacking apical pointed umbo, orange but covered dark brown at maturity, with irregular margins (figures a,b)
Stipe: 3-4 cm, yellow when young to orange or orange to red streaked when mature, smooth, humid, central, without annulus (figures b,c)
Gills: light yellow (young) to orange (mature), waxy, emarginate to free, close (figures, b,c)
Spore print: did not yield print
Microscopic
Spores: (10.5-12.0)x(7.5-7.8)µm, x̄ (11.5x7.6)µm (Qm=1.5) (n=5), elliptical to broadly oblong, hyaline (KOH), inamyloid (Melzer's), smooth, without germ-pore (figure d)
Trama: parallel (figure e: black arrow)
Cystidia: not found
Basidia: c.(30)x(7.5-9)µm, elongated and thin
Pileipellis: a dermis: of globular pigmented cells and hyphae in gelatinous substrate
Sample: GMA PT 7 PCA 5.3
4144 m   10°49'26.615" S 76°12'7.944" W
Observations: *Hygrocybe*: an epicutal dermis, and cap without an accentuated peak

1 unit = 1.5µm

1 unit = 33µm.

THE MACROFUNGI OF ANDEAN PERU   Part 1

Hygrocybe sp.. sect. Hygrocybe   Spores: none found
Southen Andes: Apurimac

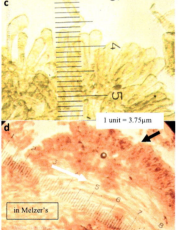

1 unit = 3.75μm

in Melzer's

Region: Ampay, Abancay, Apurimac - Southern Andes -(figure f)
Ecology: soil beside track surrounded by shrubs on edge of
Podocarpus glomeratus cloud forest (figure e)
Macroscopic
Odor: N.E.
Pileus: (2-3)cm, conical, orange slightly darker at apex (without
pointed umbo), split, smooth, moist, shiny (figure a)
Stipe: 4-5(1)cm, yellow, moist, smooth, cental, without annulus
(figure b)
Gills: white, adnexed, waxy, close to crowded
Spore print: did not yield print
Microscopic
Spores: none found
Trama: parallel, slightly dextrose (figure d: white arrow)
Cystidia: none observed
Basidia: c.(30-38)x(8)μm, piriform (figure c), dextrose (figure d: black
arrow)
Pileipellis: a cutis; of repent mycelium
Sample: GMA PT 3 PSA 2.21
3277 m   13°35'47.849" S 72°52'44.91" W
Observations: *Hygrocybe*: similar to GMA PT 7 PCA 2.10 in Ichu
grasslands Cerro de Pasco, but cap of this sample without
pronounced peak and collection environments very different.

Hygrocybe sp. sect. Hygrocybe   Spores: c.(15.0 x 9.4) μm, (Q=1.6)
Southern Andes: Cusco

Region: Zurite, Anta, Cusco - Southern Andes (figure f)
Ecology: organic material in grassland with bushes (figure f)
Macroscopic
Odor: putrid
Pileus: 5cm, narrowly conical, black (mature), smooth, moist (fig.a,b)
Stipe: 14(1cm), white with black/grey streaks when older, moist, smooth, central, broader at base, without annulus (figure b)
Gills: white, adnexed, waxy, not deliquescent, close (figure c)
Spore print: did not yield print
Microscopic
Spores: (15.0 x 9.4)μm, (Q=1.6) (n=1), broadly elliptical, hyaline (KOH), inamyloid (Melzer's), smooth, without germ-pore (figure d)
Trama: parallel (figure e:white arrow)
Cystidia: not found
Basidia: c(28-37) x(4-7)μm, elongated and thin
Pileipellis: not found (decomposed)
Samples:
GMA PT 2 PSA 6.9 (UNSAAC),
GMA PT 3 PSA 7.2
3670 m   13°26'12.048" S 72°15'10.212" W
Observations: *Hygrocybe:* characterized by carpophore shape and size combined with indication of large spore size and by old specimens that retain narrow conical caps and white gill color (c.f. other samples). Only old specimens found.

1 unit = 3.75μm

THE MACROFUNGI OF ANDEAN PERU   Part 1

P184 **_Hygrocybe sp_** sect. Hygrocybe   Spores: x̄ (7.2x4.1)μm (Qm=1.7)
Southern Andes: Cusco

Region: Ccorca, Cusco, Cusco - Southern Andes (figure c)
Ecology: in organic material below a stand of *Polylepis* trees (figure c)
Macroscopic
Odor: N.E.
Pileus: (1-3)cm, narrowly conical-campanulate, red, smooth, moist, fleshy (fig.a)
Stipe: 3-6 (1) cm, orange to red, moist, central, clavate, fibrous, without
annulus (figure a)
Gills: white, emarginate?, waxy, crowded (figure a)
Microscopic
Spores: (6.8-8.0)x(3.8-4.5)μm, x̄ (7.2x4.1)μm (Qm=1.7) (n=4), elliptical to oblong,
very thin walled, hyaline (KOH), inamyloid, possibly slightly ornamented,
withoug germ-pore (figure b)
Trama: parallel
Cystidia: not found
Basidia: with 4 spores (figure b:white arrow)
Pileipellis: N.E.
Sample: GMA PT 3 PSA 4.5
3998 m   13°33'53.88" S 72°1'41.064" W
Observations: *Hygrocybe* characteristics: carpophore shape different to others
collected. Not found in current local and regional literature.

1 unit = 3.75μm

Hygrocybe sp.   sect. Hygrocybe   Spores: x̄ (7.6x 3.9)µm (Qm=1.9)
Southern Andes: Cusco

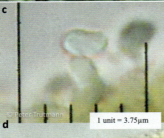

Region: Zurite, Anta and Chinchero, Cusco - Southern Andes (figure d)
Ecology: soil in pasture surrounded by *Eucalyptus* trees, or in *Polylepis* cloud forest (figure f)
Macroscopic
Odor: mild to like perspiration (body odor)
Pileus: 2cm, conical - convex finely striate with crenate margin (young),(figure a) to plane upturned (mature), (figure b), red, dry, dull
Stipe: (5-7)cm, yellow-orange, fibrous, smooth, dry, central, without annulus (figure a)
Gills: white-light yellow/orange, emarginate, waxy, distant (figure a)
Spore print: did not yield print
Microscopic
Spores: (7.5-8.3)x(3.5-4.0)µm, x̄ (7.6x 3.9)µm (Qm=1.9) (n=10), elliptical, hyaline (KOH), inamyloid (Melzer's) smooth, without germ-pore (figure c)
Trama: parallel (figure d: white arrow)
Cystidia: none found
Basidia: c. (28-34)x(7)µm with 2 or more sterigmata
Pileipellis: a cutis: carpet of mycelium
Sample:
Chinchero, Cusco GMA PT 3 PSA 5.12
3754 m   13°36'11.921" S 72°12'18.906" W
Zurite, Cusco GMA PT 3 PSA 7.24
4033 m   13°25'46.763" S 72°15'24.563" W
Observations: *Hygrocybe*: characterized and differs by carpophore shape, red color and dry cap and smaller spore size.

P186 **_Hygrocybe sp_** sect. Hygrocybe Spores: x̄ (9.7x5.6)μm (Qm=1.7)
Central Andes: Ancash

Region: Yungay, Ancash - Central Andes (figure f)
Ecology: Organic material in *Polylepis besseri* dominated
cloud forest (figure f)
Macroscopic
Odor: N.E.
Pileus: (1-2)cm, conical, orange, lightly corrugated, smooth,
dry, shiny (figures a,b)
Stipe: (5-7)cm, yellow, smooth, thin, fibrous, smooth,
central, without annulus (figure b)
Gills: white- light yellow, waxy, thick, adnate, close (figure c)
Spore print: did not yield print
Microscopic
Spores: (7.9-11.3)x(4.5-6.3)μm, x̄ (9.7x5.6)μm (Qm=1.7)
(n=5), elliptical, hyaline(KOH), inamyloid (Melzer's) smooth,
without germ-pore (figure d)
Trama: N.E.
Cystidia: not found
Basidia: c.(30x4)μm, inamyloid, elongated and narrow
Pileipellis: a cutis: repent hyphae in gelatinous medium -
scalp section (figure e: black arrow)
Sample: GMA PT 4 PNA 1.5
3800 m   9°4'46.247" S 77°39'8.993" W
Observation: *Hygrocybe:* characterized and separated by
fragile carpophore form and spore size and habitat.

1 unit = 3.75μm

***Hygrocybe* sp.**   sect. Hygrocybe   Spores: none found
Northern Andes: Lambayeque

Region: Cañaris, Ferreñafe, Lambayeque-Northern Andes (fig. h)
Ecology: organic material in relic northern Amazonian highland cloud forest (figure h)
Macroscopic
Odor: almost sweet
Pileus: (2-3)cm, in-turned convex, red, ornamented finely micaceous, moist, shiny (figures a,b, c)
Stipe: 9cm, red, with micaceous ornamentation, with slightly dark streaks, wide relative to cap, central, without annulus (figure c)
Gills: white, waxy, adnate, close-crowded (figure d)
Spore print: did not yield print
Microscopic
Spores: none found
Trama: appears parallel, dextrose (figure g; black arrow)
Cystidia: possibly tibliiform pleurocystidia with apical wartlike growths (figure f)
Basidia: c.(38-40)x(3-4) µm, immature, elongated, narrow, lightly dextrose (figure e: red arrow)
Pileipellis: not evaluated
Sample: GMA PT 3 PNA 2.10
2803 m   6°3'43.308" S 79°15'6.611" W
Observations: A *Hygrocybe:* see Singer (1975:207-11). distinguished by carpophore size, shape, and cystidia. Perhaps similarity to *H. firmus var. firmus* Plate 1 y p17 (Dennis 1970)?

1 unit = 3.75µm

in Melzer's

in Melzer's

THE MACROFUNGI OF ANDEAN PERU   Part 1

P188 ***Hygrocybe sp*** sect. Hygrocybe   Spores: x̄ (9.1x6.9)μm (Q=1.3),
Northern Andes: Lambayeque

Region: Cañaris, Ferreñafe, Lambayeque - Northern Andes (fig. h)
Ecology: poor soil on roadside embankment close to forest (fig. g)
Macroscopic
Pileus: (0.1-0.2)cm, convex, red, ornamented with glistening atomate, particles, dry, shiny (figures a,b)
Stipe: 0.5cm, orange-red, wide in relation to cap, smooth, dry, central, without annulus (figure c)
Gills: white, waxy, thick, adnate-decurrent, distant (figure c)
Spore print: too small to take
Microscopic
Spores: (7.9-10.9)x(5.2-7.5)μm, x̄ (9.1x6.9)μm (Q=1.3) (n=8), elliptical, smooth to lightly ornamented, inamyloid (Melzer's), without germ-pore (figure d)
Trama: parallel?, dextrose
Cystidia: narrow pleurocytidia and cheilocystidia between basidia? basidia? (figure d,f: black arrows) and clavatiform cheilocystidia? (figure f: yellow arrow)
Basidia: c.(48-50)x(3-4) μm, elongated, dextrose, with 4 spores (figure e: white arrow)
Pileipellis: N.E.
Sample: GMA PT 5 PNA 2.14
2360 m   6°4'33.281" S 79°12'9.143" W
Observations: *Hygrocybe:* Singer (1975:207-211), Characterized by location, its minute red, atomate carpophore, spore size and shape and possibly cystidia. Smaller than *H. miniata* in Costa Rica (Mata 1999:102).

Hygrocybe sect. *Hygrocybe*   Spores: x̄ (7.6x6.3)μm (Qm=1.2)
Northern Andes: Cajamarca

Region: Colasay, Jaen, Cajamarca - Northern Andes (figure f)
Ecology: organic matter below primary relic northern Amazonian highland cloud forest (figure e)
Macroscopic
Odor: like sewage (old sample)
Pileus: 5cm, conical, black, shiny, smooth, moist (figures a,b)
Stipe: (1-2)cm, yellow to black, smooth, fibrous moist, without annulus (figures b,c)
Gills: white turning black with age, waxy, adnate-sub-decurrent, close (figure c)
Spores: did not yield print
Microscopic
Spores: (6.3-8.3)x(4.5-7.0)μm, x̄ (7.6x6.3)μm (Qm=1.2) (n=6), elliptical -subglobose (pea form), appear smooth or slightly ornamented, hyaline(KOH), inamyloid (Melzer's), without germ-pore (figure d)
Trama: not found
Cystidia: not found
Basidia: not found
Pileipellis: not found
Sample: GMA PT 5 PNA 3.13
est. 2435 m   5°58'11.29" S 79°6'17.64" W
Observations: *Hygrocybe*: Singer (1975) p.207. Old sample with maggots. Characterized by spore size and shape as well as carpophore shape. Possibly *H. conicus var. peradenyca* (Dennis 1970:17), with yellow-black basidiophore but its spores appear larger (7-10)x(5-6)μm. *H.conica* also reported in Costa Rica (Mata 1999:98) and Colombia (Franco-Molano et al 2005:68). Younger samples needed.

1 unit = 3.75μm

Hygrocybe s.l.

P190   **Hygrocybe** *s.l.*   *Spores:* none found
Northern Andes: Lambayeque

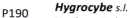

in Melzer's

1 unit = 3.75µm

Region: Cañaris, Ferreñafe, Lambayeque - Northern Andes (fig. d)
Ecology: on dead wood with moss in northern relic Amazonian
cloud forest (figure d)
Macroscopic
Pileus: 0.5cm, convex, orange, viscid, smooth (figure a)
Stipe: 3cm, orange, smooth, fragile, central, without annulus (fig. a)
Gills: orange, sub-decurrent, distant?
Spore print: too small
Microscopic
Spores: none found
Trama of the hymenophore; not found
Cystidia: not found
Basidia: not found
Pileipellis: a cutis (enterocutis?): of repent inflated hyphae(black
arrow) in gelatinous medium - radial section (figures b) scalp
section (figure c)
Sample: GMA PT 3 PNA 2.24
2727 m   6°3'38.94" S 79°14'59.177" W
Observations: tentatively *Hygrocybe s.l.*: it resembles
Glioxanthomyces. Immature specimen. .

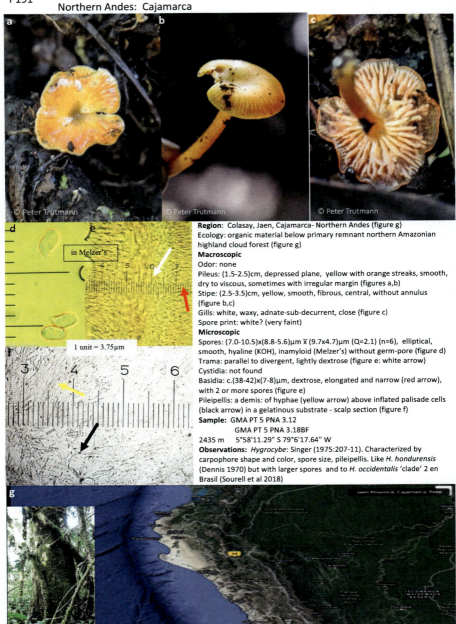

Hygrocybe s.l.
P191
Northern Andes: Cajamarca

Spores: x̄ (9.7x4.7)µm (Q=2.1)

in Melzer's

1 unit = 3.75µm

Region: Colasay, Jaen, Cajamarca- Northern Andes (figure g)
Ecology: organic material below primary remnant northern Amazonian highland cloud forest (figure g)
Macroscopic
Odor: none
Pileus: (1.5-2.5)cm, depressed plane, yellow with orange streaks, smooth, dry to viscous, sometimes with irregular margin (figures a,b)
Stipe: (2.5-3.5)cm, yellow, smooth, fibrous, central, without annulus (figure b,c)
Gills: white, waxy, adnate-sub-decurrent, close (figure c)
Spore print: white? (very faint)
Microscopic
Spores: (7.0-10.5)x(8.8-5.6)µm x̄ (9.7x4.7)µm (Q=2.1) (n=6), elliptical, smooth, hyaline (KOH), inamyloid (Melzer's) without germ-pore (figure d)
Trama: parallel to divergent, lightly dextrose (figure e: white arrow)
Cystidia: not found
Basidia: c.(38-42)x(7-8)µm, dextrose, elongated and narrow (red arrow), with 2 or more spores (figure e)
Pileipellis: a demis: of hyphae (yellow arrow) above inflated palisade cells (black arrow) in a gelatinous substrate - scalp section (figure f)
Sample: GMA PT 5 PNA 3.12
        GMA PT 5 PNA 3.18BF
2435 m   5°58'11.29" S 79°6'17.64" W
Observations: *Hygrocybe*: Singer (1975:207-11). Characterized by carpophore shape and color, spore size, pileipellis. Like *H. hondurensis* (Dennis 1970) but with larger spores and to *H. occidentalis* 'clade' 2 en Brasil (Sourell et al 2018)

Hygrocybe s.l.   Not processed microscopically
Southern Andes: Cusco

Region: Zurite, Anta, Cusco - Southern Andes (figure e)
Ecology: organic material in *Polylepis besseri* cloud forest
Macroscopic
Pileus: 3cm, conical, dark purple when young to orange or yellow when mature, viscid, smooth (figs, a,b,c)
Stipe: 4 cm, yellow to light orange, smooth, central, viscid without annulus (figures a,b,c)
Gills: white, thick, adnexed to emarginate, close (figure c)
Spore print: did not yield print
Samples: GMA PT 2 PSA 6.7   UNSAAC ,
4991 m   13°25'50.142" S 72°15'22.632" W
              GMA PT 2 PSA 6.6 UNSAAC
4909 m   13°25'56.771" S 72°15'16.818" W
Observations: Conforms to a *Hygrocybe*. Not evaluated further.

Region: Putina, Puno - Southern Andes (figure d)
Ecology: soil in highland Ichu puna grassland (figure d)
Macroscopic
Pileus: (1-2)cm, conical, without accentuated central umbo, orange-blackening towards center, smooth, shiny, dry (figures, a,c)
Stipe: (3-5)cm, yellow-orange, darkening with age, with fine hairlike protrusions, central, without annulus (figures a,b)
Gills: white, waxy, lightly serrated, unusual emarginate to sub-decurrent, close (figure b)
Spore print: did not yield print
Microscopic
Not yet evaluated
Sample:
GMA PT 2 PSA 9B7
3855 m   14°58'31.164" S 69°50'42.857" W
Observations: *Hygrocybe*: carpophore characterized by color, size and pileus without accentuated umbo peak. The sample was in poor condition.

P194F **Hygrocybe s.l..** sect. *Hygrocybe*   Photo record only
Southern Andes: Puno

Region: Huancane, Puno - Southern Andes (figure c)
Ecology: soil in highland Ichu puna grassland (figure c)
Macroscopic
Pileus: (1-4)cm, conical, campanulate to depressed, dull orange, smooth, humid to dry (figures a,b)
Stipe: (3-4)cm, orange, smooth, central, without annulus, (figure a)
Gills: white to light orange, waxy, adnexed, close
Photographic sample only:
GMA PT 3 PSA 8A.11F
3827 m   15°10'46" S 69°50'38.62" W
Observations: *Hygrocybe:* characterized by basidiophore shape, small size, and dull orange color.

Hygrocybe _s.l._ (_Humidicutis?_)   Photo record only
Southern Andes: Ayacucho

Region: Lucanas, Ayacucho - Southern Andes (figure d)
Ecology: Growing on Ichu grass in highlands puna grasslands (figure d)
Macroscopic
Pileus: 1.5cm, broadly conical, bright orange, smooth, humid -dry (figure a)
Stipe: 4cm, orange, fine, fibrous, smooth, central, without annulus (fig. a,b)
Gills: white to light yellow, waxy, sub-decurrent to decurrent, subdistant (figures b,c)
Photographic sample: GMA PT 2 PSA 2.2BF
3991m   13°33'55.242" S 72°1'40.314" W
Observation: Tentatively Hygrocybe s.l., but carpophore possesses _Humidicutis_ like characters conical cap, and not deeply decurrent gills?, fragile nature, bright yellow color. Needs to be collected and checked for absence of clamp connections, and pigment that does not dissolve in alkali solutions as well as spore size that should be > 7μm (Singer 1975:211).

THE MACROFUNGI OF ANDEAN PERU Part 1

Hygrophorus

P196 ***Hygrophorus sp.*** Spores: x̄ (7.8x 4.0)μm (Qm=1.9)
Southern Andes: Cusco

Region: Zurite, Anta, Cusco - Southern Andes (figure h)
Ecology: organic material under a *Polylepis spp.* cloud forest (figure g)
Macroscopic
Pileus (1-6)cm, convex with corrugated margin, light yellow when young (figure a), to broadly convex and white when older (figure b), dry, smooth, dull to silky.
Stipe: (4-10)cm, yellow, with scales when young (figure a) to white without evident scales when older, central (figure b) without annulus (figure b)
Gills: orange, sub-decurrent, crowded (figures a,b)
Spore print: white
Microscopic
spores: (7.5-8.5)x(3.8-4.5)μm, x̄ (7.8x 4.0)μm (Qm=1.9) (n=5), elliptical to oblong, hyaline (KOH), inamyloid (Melzer's), smooth, without germ-pore (figure c)
Trama: divergent (figure e: white arrow)
Cystidia: not found
Basidia: c.(23 x 8)μm, pseudoamyloid, with 2-4 spores (fig. e)
Pileipellis: a cutis or trichodermium: of inflated repent parallel hyphae - scalp section (figure f: black arrow)
Sample: GMA PT 3 PSA 7.18
4034 m   13°25'48.293" S 72°15'24.479" W
Observation: *Hygrophorus:* due to bilateral trama. The Genus is mycorrhizal and has not been found beyond the Quercus area (N.Peru) in South America (Singer 1975:197-8).

1 unit = 3.75μm

Arrhenia

P197
(2)

Arrhenia sp.   Spores: x̄ (9.4x6.6)µm (Qm=1.4)
Southern Andes: Cusco

Region: Cusco and Tipón, Cusco - Southern Andes (figure h)
Ecology: on stone walls with moss (figure g)
Macroscopic
Pileus: (0.5-1)cm, omphaloid, deeply depressed with in-turned undulating margins, grey-brown, smooth, metallic (figs. a,b)
Stipe: (1-2)cm brown-grey, slender, tough, smooth, central, without annulus (figure b)
Gills: grey, wide, distant, decurrent (figure b)
Spore print: did not yield print
Microscopic
Spores: (9-10)x(6-7)µm, x̄ (9.4x6.6)µm (Qm=1.4) (n=5), subglobose, hyaline (KOH), inamyloid (Melzer's) smooth with inclusions, or possibly slightly ornamented, without germ-pore (figure c)
Trama: of parenchyma like cells (black arrow), without clamp connections (figure e)
Cystidia: not found
Basidia: with 4 spores (figure d: red arrow)
Pileipellis: a cutis: of pigmented repent hyphae (white arrow) with non-gelatinous cellular subdermis-scalp section (fig. f:)
Sample:
Cusco, Cusco GMA PT 1 PSA 4.2
3386m   13°31'16.27S 71°58'23.78 W
Tipón, Cusco: GMA PT 1 PSA 5.13a
3448 m   13°34'16.23" S 71°47'5.225" W
Observation: *Arrhenia: see* (Laessoe T. and Petersen J.H., 2008:129)

1 unit = 3.75µm

1 unit = 3.75µm

Cora

Cora sp.   Photographic record only
Southern Andes: Cusco and Puno

Region: Canchis, Cusco and Huancane, Puno - Southern Andes (figure d)

Ecology: on soil and rock associated with branched bryophytes and moss in highland puna (fig. e)

Macroscopic

General: (20-30)cm across, foliose, flat, composed of 10-30 or more semicircular, imbricate lobes; individual lobes (2-3 x 3-6)cm (figures a, b)

Thallus upper side: light and dark grey, thin, leaf-like, slightly undulate, and zoned, with distinct light, often bent, margins (figure c)

Thallus lower side: light grey

Photographic samples:

Canchis, Cusco: GMA PT 1 PSA 3.5F
3422 m   13°58'20.994" S 71°29'34.781" W
Huancane, Puno: GMA PT 1 PSA 10.9BF
3883 m   15°10'34.594" S 69°50'41.532" W

Observations: *Cora*: similar to *Cora squamiformis* found in Colombia (Lücking et al 2013).

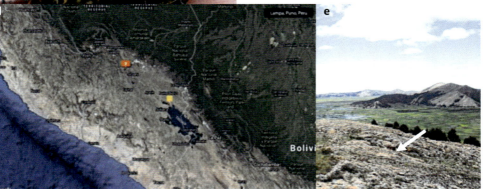

P199F

Cora *cf.* ***cyphellifera***   Photographic record only
Northern Andes: Lambayeque

Region: Cañaris, Ferreñafe, Lambayeque - Northern Andes (figure c)
Ecology: on wood with branched bryophytes in relic northern Amazonian highland cloud forest (figure d)
Macroscopic
General: around 20cm across, foliose, flat, composed of 10-20 semicircular, imbricate lobes; individual lobes (5-15 x 5-10)cm (figures a, b)
Thallus upper side: blue-green, thin, leaf-like, with pitted surface, slightly undulate, and zoned (figure b)
Thallus lower side: blue-green with peeling white epicutis layer (figure b: white arrow)
Photographic sample:
GMA PT 3 PNA 2.12BF
2725 m   6°3'42.179" S 79°15'7.457" W
Observations: *Cora:* fits description of *Cora cyphellifera* Dal-Forno, Bungartz and Lücking described in Ecuador (Lücking et al., 2013)

Pseudoarmillariella

Pseudoarmillariella sp.?

P200

Central Andes: Ancash

Spores: x̄ (5.6-3.9) µm (Qm=1.4)

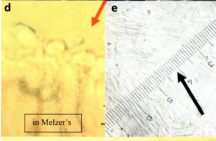

in Melzer's

1 unit = 3.75µm

in Melzer's

Region: Recuay, Ancash - Central Andes (figure g)
Ecology: organic material in native forest by river (figure g)
Macroscopic
Odor: like fresh soil (fresh), like champignons (dry)
Pileus: (1.5-3.5)cm, omphalinoid depressed with in-turned undulating margins, cream, hygrophanous, smooth, dull (figs. a,b)
Stipe: (2.5-3.5)cm, white to cream, slender, tough, smooth, without annulus (figure c)
Gills: off white-cream, lightly serrated, sub-decurrent, close (figure c)
Spore print: light cream
Microscopic
Spores: (5.6)x(3.7-4.0)µm, x̄ (5.6-3.9)µm (Qm=1.4) n=(2), few, elliptical to ovate, hyaline (KOH), irregularly inamyloid and lightly amyloid (Melzer's), smooth, without germ-pore (figure d)
Trama: regular, inamyloid (figure f: white arrow)
Cystidia: not found
Basidia: with 2 or more spores, lightly dextrose, with inconsistent amyloid reactions around the sterigmata y spores (fig.f:red arrow)
Pileipellis: a cutis: of prostrate non-specialized hyphae - scalp section (figure e: black arrow)
Sample: GMA PT 3 PNA 9.23
1658 m   9°3'52.476" S 77°38'7.367" W
Observactions: *Pseudoarmillariella:* (Singer 1975) p.283: omphaloid habit, lignicolous, laminae usually yellow or yellowish not strongly forked, spores oblong-ellipsoid and amyloid, pigments incrusting, pileus hygrophanous sub punctulate, no cystidia.

Lichenomphalia

P201F   ***Lichenomphalia?***   Photographic record only
         Southern Andes: Puno

Region: Melgar, Puno - Southern Andes (figure c)
Ecology: soil with moss, algae and grass in puna grassland (figure c)
Macroscopic
Pileus: (0.5-1.0)cm, omphaloid, deeply depressed plane with uneven eroded margins, yellow, ridged, smooth, delicate, (figures a,b)
Stipe: (0.5-1)cm, cream to yellow, smooth, without annulus (fig. c)
Gills: shallow, yellow, decurrent, distant (figure c)
Spore print: no
Sample: GMA PT 1 PSA 7.6
4006 m   14°38'36.227" S 70°44'44.376" W
         GMA PT 1 PSA 7.10
3984 m   38'40.193" S 70°44'51.149" W
Observations: Like *Lichenomphalia*: see Laessoe and Petersen and Petersen (2019)

Cuphophylloid Grade

Cuphophyllus

P202   *cf.* **Cuphophyllus sp.**   Spores: none found
      Southern Andes: Cusco

Region: Ccorca, Cusco, Cusco - Southern Andes (figure g)
Ecology: associated with Ichu in highland puna (figure f)
Macroscopic
Odor: N.E.
Pileus 2cm, plane (flat with deep margins), white, smooth, dull (figure a)
Stipe:(3-5)cm, light yellow, central, smooth, hollow, without annulus (figure b)
Gills: yellow-orange, waxy, sub-decurrent, crowded (fig. b)
Spore print did not yield a print
Microscopic
Spores: not found
Trama: irregular-interwoven, inamyloid (figure d)
Cystidia: none found
Basidia: c.(40x4)μm, elongated, narrow, inamyloid (figure c: red arrow)
Pileipellis: a cutis: of interwoven undifferentiated mycelium -scalp section (figure e: black arrow)
Sample: *GMA PT 3 PSA 4.4
3984 m   13°33'53.532" S 72°1'41.712" W
Observations: *Like Cuphophyllum*: due to its interwoven hymenophore like trama and other characteristics such as small size and short rather thin, short decurrent gills (Laessoe and Petersen, 2019: 143) *sample lost

in Melzer's

1 unit = 3.75μm

P203   *Cuphophyllus sp.*   Spores: x̄ (7.6x4.2)µm (Qm=1.8)
Southern Andes: Cusco

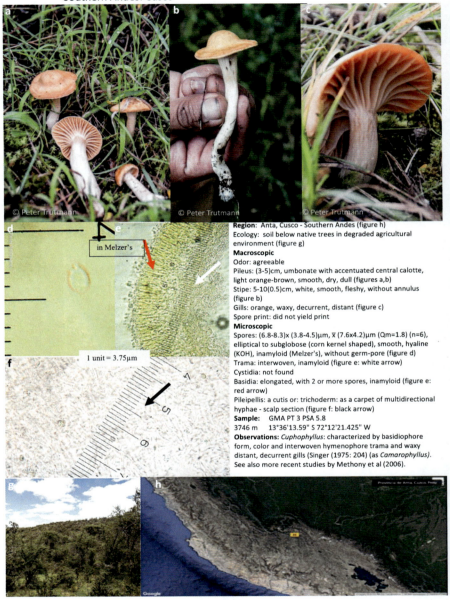

© Peter Trutmann

in Melzer's

1 unit = 3.75µm

Region: Anta, Cusco - Southern Andes (figure h)
Ecology: soil below native trees in degraded agricultural environment (figure g)
Macroscopic
Odor: agreeable
Pileus: (3-5)cm, umbonate with accentuated central calotte, light orange-brown, smooth, dry, dull (figures a,b)
Stipe: 5-10(0.5)cm, white, smooth, fleshy, without annulus (figure b)
Gills: orange, waxy, decurrent, distant (figure c)
Spore print: did not yield print
Microscopic
Spores: (6.8-8.3)x (3.8-4.5)µm, x̄ (7.6x4.2)µm (Qm=1.8) (n=6), elliptical to subglobose (corn kernel shaped), smooth, hyaline (KOH), inamyloid (Melzer's), without germ-pore (figure d)
Trama: interwoven, inamyloid (figure e: white arrow)
Cystidia: not found
Basidia: elongated, with 2 or more spores, inamyloid (figure e: red arrow)
Pileipellis: a cutis or: trichoderm: as a carpet of multidirectional hyphae - scalp section (figure f: black arrow)
Sample: GMA PT 3 PSA 5.8
3746 m   13°36'13.59" S 72°12'21.425" W
Observations: *Cuphophyllus*: characterized by basidiophore form, color and interwoven hymenophore trama and waxy distant, decurrent gills (Singer (1975): 204) (as *Camarophyllus*). See also more recent studies by Methony et al (2006).

P204 **_Cuphophyllus pratensis_** Not evaluated microscopically
Southern Andes: Cusco

Region: UNSAAC station, Cusco -Southern Andes (figure d)
Ecology: soil and Ichu in highland puna grasslands (fig. d)
Macroscopic
Odor: agreeable
Pileus: 2-3 (0.2)cm, broadly convex when young to slightly
depressed plane, orange- cream, smooth, rimose like
cracks (figs. a,b)
Stipe: 3-5(0.5)cm, white, smooth, central to slightly off-
center, without annulus (figure c)
Gills: white to cream, decurrent, waxy, wide, distant (fig. c)
Spore print: white
Microscopic
Not evaluated
Sample: GMA PT 2 PSA 5.16
3593 m 13°34'20.645" S 71°51'20.862" W
Observations: _Cuphophyllus_: similar to C. _pratensis_ (Babos
et al., 2011). Sample is stored at Universidad Nacional San
Antonio Abad de Cusco (UNSAAC). Not _Clitocybe_ due to
waxy wide, distant gills.

Cuphophyllus sp
Central Andes: Ancash

Spores: x̄ (7.9x4.5) μm (Q=1.7)

© Peter Trutmann   © Peter Trutmann   © Peter Trutmann

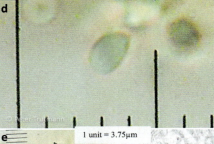

© Peter Trutmann

1 unit = 3.75μm

in Melzer's

Region: Yungay, Ancash - Central Andes (figure g)
Ecology: organic material in *Polylepis* spp. dominated cloud forest (figure g)
Macroscopic
Odor: dog odor
Pileus: 2.5cm, lightly depressed plane, in rolled, yellow-brown, smooth, sometimes lacinate at center, dry, dull (figure a)
Stipe: (8-10)cm, white, smooth, central, without annulus (figure b)
Gills: white, decurrent, waxy, wide, distant (figures b,c)
Spore print: did not yield print
Microscopic
Spores: (8.2)x(4.8) μm, x̄ (7.9x4.5)μm (Q=1.7) (n=2), scarce, elliptical, hyaline (KOH), inamyloid (Melzer's), smooth (figure d)
Trama: interwoven (figure e: white arrow)
Cystidia: not found
Basidia: c.(22-5-26)μm, inamyloid (figure e)
Pileipellis: a cutis or trichoderm: a carpet of hyphae-scalp section (figure f: black arrow)
Sample: GMA PT 3 PNA 8.16
3746m   9°5'6.066" S 77°39'33.9" W
Observations: *Cuphophyllus*: (see Singer 1975:204). Characterized from other samples by having depressed pileus as well as carpophore shape and size.

© Peter Trutmann

Cuphophyllus sp. Spores: x̄ (3.3x3.3)μm (Qm=1.0)
Northern Andes: Cajamarca

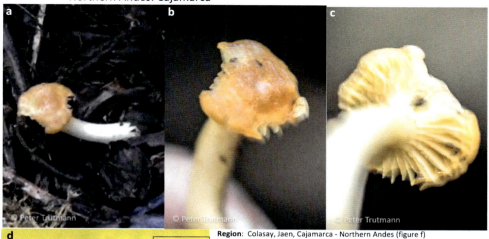

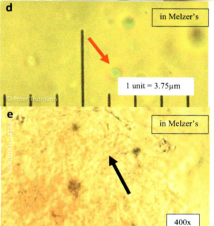

in Melzer's

1 unit = 3.75μm

in Melzer's

400x

Region: Colasay, Jaen, Cajamarca - Northern Andes (figure f)
Ecology: organic material below relic Amazonian cloud forest (figure g)
Macroscopic
Odor: N.E.
Pileus: 0.5cm, broadly mammilate-umbonate with large calotte, orange brown, smooth, shiny (figs.a,b)
Stipe: 2cm, white, smooth, central, without annulus (figure a,c)
Gills: white, wide, decurrent, distant (figure c)
Spore print: did not yield print (sample in poor condition)
Microscopic
Spores: (3.0-3.8)x(3.0-3.8)μm, x̄ (3.3x3.3)μm (Qm=1.0) (n=6), very small, globose to subglobose, hyaline (KOH), inamyloid (Melzer's), smooth (fig. d)
Trama: uncertain - tissue very degraded
Cystidia: not found
Basidia: not found
Pileipellis: a cutis or trichoderm: of repent hyphae in a gelatinous medium - scalp section (figure e:black arrow)
Sample: GMA PT 5 PNA 3.11
2435 m   5°58'11.29" S 79°6'17.64" W
Observations: *Cuphophyllus:* mainly due to carpophore form Singer (1975:204). Spores were very difficult to visualize clearly possibly because of the degraded nature of the sample. However, the possibility exists that they were micro-basidiospores from dimorphic basidia as has been found for the tropical *Hygrocybe firma (Lodge et al., 2014),* or not spores at all.

Cuphophyllus?
Northern Andes: Cajamarca

Spores: x̄ (6.8x4.2)μm (Qm=1.6)

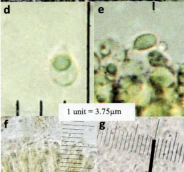

1 unit = 3.75μm

Region: Cumbemayo, Cajamarca, Cajamarca -Nothern Andes (figure h)
Ecology: on soil in puna grassland (figure h)
Macroscopic
Odor: like radish
Pileus: (2.5-3.5)cm, broadly convex, cream to light brown, smooth, bright, dry, hygrophanous (figures a,b)
Stipe: 3-4(0.5)cm, white, smooth, centric to acentric, without annulus (figure c)
Gills: white to off-white, thick and waxy, decurrent, close (figure c)
Spore print: white (figure c: orange arrow)
Microscopic
Spores: (5.6-7.5)x(3.7-4.5)μm, x̄ (6.8x4.2)μm (Qm=1.6) (n=5), elliptical to corn shaped, hyaline (KOH) inamyloid (Melzer's), smooth, without germ-pore (figure d)
Trama: uncertain, inamyloid
Cystidia: not found
Basidia: c.(50-55)x(4)μm, elongated with 2 or more sterigmata (figure f): red arrow)
Pileipellis: a cutis?: of enlarged, interwoven hyphae in a gelatinous matrix - scalp section (figure g: black arrow)
Sample: GMA PT 3 PNA 4.15
3565 m   7°11'24.941" S 78°34'40.487" W
Observations: Like a *Cuphophyllus*: Singer (1975:204)(as *Camarophyllus*). The hymenophore trama structure needs verification.

P208 *cf.* **Cuphophyllus**   Spores: x̄ (7.2x3.9) µm (Qm=1.8)
Central Andes: Ancash

Region: Llanganuco, Yungay, Ancash - Central Andes (figure g)
Ecology: soil in native pasture by lake Chinancocha and *Polyleptis* cloud forest (figure g)
Macroscopic
Odor: mild (fresh), agreeably strong (dry)
Pileus: (3-4.5)cm, depressed plane, off-white to cream, dry, dull, smooth, with slightly hygrophanous margins when old (figures a,b),
Stipe: (3-4)cm, white-cream, fibrous, smooth, central, without annulus (fig. c)
Gills: white, wide, waxy, adnexed to emarginate, close
Spore print: white (figure c: orange arrow)
Microscopic
Spores: (6.2-7.5)x(3.8-4.2)µm, x̄ (7.2x3.9)µm (Qm=1.8) (n=5), small, elliptical to pip-shaped, thin walled, hyaline (KOH), inamyloid (Melzer's), smooth, without germ-pore (figure d)
Trama: +/-interwoven (figure e: white arrow)
Cystidia: not found
Basidia: c.(38x4)µm, elongated and narrow (figure e)
Pileipellis: a cutis (enterocutis?): of interwoven hyphae - scalp section (figure)
Sample: GMA PT 5 PNA 1.18
3854 m   9°3'50.074" S 77°38'1.866" W
Observations: Like *Cuphophyllus,* with dull colored in amyloid spores and trama is interwoven (Arora 1987, Singer 1975), but like *Neohygrophorus* with thick, greasy, emarginate (not decurrent) gills (Laessoe and Petersen, 2019:143)

1 unit = 3.75µm

Family Hymenogastraceae

© Peter Trutmann

Galerina

Galerina sp.   Spores: x̄ (8.3x4.8)μm (Qm=1.7)
Southern Andes: Apurimac

Region: Abancay, Apurimac - Southern Andes (figure h)
Ecology: in moss below native *Podocarpus glomeratus* cloud forest and shrubs (figure g)
Macroscopic
Odor: N.E.
Pileus: (1-2)cm, conical to broadly convex, cream - yellow, tacky - viscid, smooth, translucent striate (figures a,b)
Stipe: 3cm, white to brown, smooth, central, without annulus (figures b,c)
Gills: cream to light brown, adnexed to sub-decurrent, close to crowded (figure c)
Spore print: brown
Microscopic
Spores: (8.2-8.6) x (4.5-4.9) μm, x̄ (8.3x4.8)μm (Qm=1.7) (n=5), elliptical, ornamented, with a plaque (red arrow), brown (KOH), without germ-pore (figures d,f)
Cystidia: not found
Basidia: with 4 spores (figure f: white arrow)
Pileipellis: a cutis: a carpet of repent parallel thin undifferentiated hyphae- scalp section (fig. e: black arrow)
Sample: GMA PT 3 PSA 2.15
3190 m   13°35'51.701" S 72°52'43.037" W
Observations: *Galerina*: by its brown spores with plaques as well as other basidiophore characters (Singer 1975:630).

1 unit = 3.75μm

Galerina sp.

Central Andes: Lima

Spores: x̄ (11.7x6.9) µm (Qm=1.7)

© Peter Trutmann

1 unit = 1.5µm

1 unit = 3.75µm

1 unit = 6.5µm

Region: Zarate, Huarochiri, Lima - Central Andes (figure h)
Ecology: leaf litter below native shrubs (figure h)
Macroscopic
Odor: N.E.
Pileus: (1.5-2)cm, collyboid, plane slightly depressed, orange, striate, smooth, dry, silky, slightly corrugated at margin (figs. a,b)
Stipe: (2-3)cm, central, thin, orange, smooth without annulus (figures b,c)
Gills: orange, adnate, relatively superficial, close
Spore print: did not yield print
Microscopic
Spores: (9.8-13.3)x(6.0-7.5)µm, x̄ (11.7x6.9)µm (Qm=1.7) (n=5), elliptical to amygdaliform, brown-deep rusty brown (KOH), smooth, with plaque, without germ-pores (figure d)
Trama: irregular (subregular?) (figures f,g)
Cystidia: metuloid cheilocystidia (54-60x12)µm (figure g: orange arrow) and pleurocystidia (54-60x 12)µm (figure f: red arrow)
Basidia: c. (30x7)µm with 2-4 spores (figure d: white arrows)
Pileipellis: a cutis: an interwoven carpet of repent, enlarged hyphae - scalp section (figure e: black arrow)
Sample: GMA PT 6 PCA 1.6
2869 m 11°55'43.62" S 76°29'6.407" W
Observations: *Galerina*: spores with plaques, spore color, adnate gills, pileipellis form of repent hyphae and carpophore habit (see Singer (1975:629), Dennis (1970:272), Arora (1986:399). (n.b.*Tubaria* has decurrent gills)

© Peter Trutmann

Gymnopilus

Gymnopilus *cf.* ***pratensis*** Spores: x̄ (7.3x4.6)µm (Qm=1.6)
Southern Andes: Puno

Region: Lampa, Puno -Southern Andes (figure i)
Ecology: in organic material in highland agro-grassland (fig.h)
Macroscopic
Color reaction: none
Pileus: (2-3)cm, convex to broadly convex, orange, smooth, dry
(figures a,b)
Stipe: 2-3(0.5)cm, fleshy, short, covered with a white cottony
mycelial layer above orange, centric, without annulus (figs. a,b)
Gills: orange, adnexed, crowded (figure c)
Spore print: orange (orange arrow)
Microscopic
Spores: (7-8)x(4-5)µm, x̄ (7.3x4.6)µm (Qm=1.6) (n=5), elliptical,
orange (KOH), ornamented, without 'plaque', generally without
but rarely with narrow germ-pore (figures d,e)
Trama: made of palisade like inflated cells (fig. e: black arrow)
Cystidia: not found
Basidia: with 2-4 spores
Pileipellis: mycelial, of inflated lightly pigmented hyphae (fig. g)
Sample: GMA PT 1 PSA 1.1
3845 m   15°18'46.499" S 70°12'42.978" W
Observation: *Gymnopilus*: because of carpophore and spore print
color, generally germ-pore-less, ornamented plaqueless spores,
and filamentous pileipellis Dennis (1970:407)

P163 **Gymnopilus sp.** Spores: x̄ (11.1x6.6)μm (Qm=1.7)
Southern Andes: Arequipa

Region: Arequipa, Arequipa - Southern Andes (figure g)
Ecology: soil associated with Ichu grass in highland sandy puna grasslands at a site with archaeological remains of ceramics (figure b,f)
Macroscopic
Odor: strong
Pileus: (3-10)cm, broadly convex, incurved, orange, fleshy, with light margin, with light to dark orange fine hairlike scales (fig. a)
Stipe: 8-15(1)cm, fleshy, covered with a white cottony mycelial layer above an orange base, central, with fibrous annulus (figure c)
Gills: white to orange, adnate, crowded (figure c)
Spore print: rusty orange
Microscopic
Spores: (9.5-11.6)x(5.6-7.5)μm, x̄ (11.1x6.6)μm (Qm=1.7) (n=5), elliptical, orange (KOH), ornamented, without plaque, and without germ-pore (figures d,e)
Sample:
GMA PT 2 PSA 10.15,
GMA PT 2 PSA 10.17,
GMA PT 2 PSA 10.10
3967 m   16°3'16.302" S 71°20'55.674" W
Observations: *Gymnopilus*: by carpophore, spore color and shape. Similar to, but darker than, lignicolous *G.spectabilis* with spores (9-10)x(6.5-7)μm, (Dennis 1970:411)

1 unit = 3.75 μm

Psilocybe

cf. ***Psilocybe sp.***   Spores: x̄ (16.6x9.8)µm (Qm=1.7)
Southern Andes: Puno

10 µm

Region: Melgar, Puno - Southern Andes (figure g)
Ecology: soil and Ichu grass litter in altiplano puna grasslands (fig.f)
Macroscopic
Odor: N.E.
Color reaction: brown to bluish
Pileus: (1- 1.5)cm, convex, black when mature, moist-dry, shiny, striated, hygrophanous? (figures a,b)
Stipe: (3-4)cm, brown, smooth, central, farinose by cap, without annulus (figure b)
Gills: black, spotted, emarginate, close (figure b)
Spore print: purplish brown
Microscopic
Spores: (16-17)x(9-11)µm, x̄ (16.6x9.8)µm (Qm=1.7) (n=5), elliptical to slightly citriform, thick walled, dark brown (KOH), smooth, with prominent germ-pore (figure c)
Trama: regular (figure d: red arrow),
Cystidia: unclear (no yellowing chrysocystidia)
Basidia: c.(28 x10)µm, broad, with 4 spores (flecha blanca)
Pileipellis: a cutis (enterocutis ?): of relatively parallel, repent, inflated hyphae -scapl section (figure e: black arrow)
Samples:
GMA PT 1 PSA 7.7
GMA PT 1 PSA 7.12
3972 m   14°38'41.981" S 70°44'52.421" W
Observations: tentatively in *Psilocybe: if reaction blue. A Deconica if not:* spore print purplish brown, pileipellis a cutis, bluish reaction?, and no yellowing chrysocystidia or pleudocystidia found (Dennis 1970: 70). Old sample.

10 µm   10 µm

P212   *cf.* **Psilocybe sp.**   Spores: x̄ (13.3x8.8)μm (Qm=1.5)
Southern Andes: Puno

Region: Melgar, Puno Southern Andes (figure h)
Ecology: cow-dung in altiplano puna grasslands (figure e)
Macroscopic
Color reaction: bluish
Pileus: 2cm, campanulate, purple-brown, moist, lightly
rimulose, with slightly eroded margin (figure b: top 4)
Stipe: 3cm, brown, central, without annulus (figure b: top 4)
Gills: light purple, adnexed, close
Spore print: dark purple
Microscopic
Spores: (12-14)x(8-9)μm, x̄ (13.3x8.8)μm (Qm=1.5) (n=5),
eliptical, thick walled, brown(KOH), smooth, with germ-pore
(figure c)
Trama: interwoven
Cystidia: not found
Basidia: short with 4 spores
Pilepellis: a cutis: of repent interwoven hyphae - scalp section
(figure d: black arrow)
Sample: GMA PT 1 PSA 7.8
3992 m   14°38'37.685" S 70°44'47.159" W
Observations: *Psilocybe*: bluish reaction, an epicutis as a cutis,
dark purple spore print and lack of chrysocystidia (Dennis
1970:70)

10 μm

10 μm

Psilocybe cubensis

P213

Central Andes: Ancash

Spores: (x̄ (14.6x8.7 µm (Qm=1.7)

a
b
c

© Peter Trutmann   © Peter Trutmann   © Peter Trutmann

d
e
1 unit = 3.75µm
1 unit = 6.5µm

f   g   h
1 unit = 3.75µm

Region: Llanganuco, Yungay, Ancash - Central Andes (figure i)
Ecology: on cow-dung in a *Polylepis* spp. dominated cloud forest (figure i)
Macroscopic
Odor: agreeable
Color reaction: no reaction
Pileus: 3-10(0.5)cm, conical to broadly convex to plane with an umbo, cream to golden brown, smooth to viscid, shiny, hygrophanous (figures a,b)
Stipe: (5-15)cm, white to yellowish cream, smooth or lightly scaled, with a prominent annulus (figures b,c)
Gills: light purple darkening with age, slightly mottled, adnexed, close (fig. c)
Spore print: purple brown
Microscopic
Spores: (14.2-15.4)x(7.9-9.5)µm, x̄ (14.6x8.7)µm (Qm=1.7) (n=6), truncated elliptical, thick walled, brown (KOH), smooth with prominent germ-pore (figure d)
Trama: N.E.
Cystidia: yellow chrysocystidia (KOH) as pleurocysitidia (figure f: white arrow), and as cheilocysitidia (figure g: yellow arrow)
Basidia: with 2-4 sterigmata (figure h: red arrow)
Pileipellis: a cutis: of repent undifferentiated hyphae - scalp section (figure e: black arrow)
Sample: GMA PT 4 PNA 1.1
3843 m   9°4'44.724" S 77°39'9.384" W
Observations: *Psilocybe:* there was no blue reaction, but it's macro and microscopic characteristics are those classically *Psilocybe cubensis* (Stamets P., 1996:108).

i

Family Inocybaceae

© Peter Trutmann

Inocybe

Inocybe sp.
Central Andes: Ancash

Spores: x̄ (10.5x5.6) µm (Q=1.9)

Region: Yungay, Ancash - Southern Andes (figure i)
Ecology: soil in ichu based altiplano puna grassland (figure h)
Macroscopic
Odor: unpleasantly sour
Pileus: 2cm, umbonate, light darkening bright brown and dark brown at the umbo, dry, silky, without scales, (figure a)
Stipe: 4cm, white-cream, covered by moist membranous tissue, otherwise smooth, with remnants of a fibrous annulus (figure b)
GIlls: white to dull clay colored, adnate, close to crowded (figure c)
Spore print: a dull darkish brown (figure c: orange arrow)
Microscopic
Spores: (9.0x10.9)x(5.6)µm, x̄ (10.5x5.6) µm (Qm=1.9) (n=5), elliptical, smooth, brown(KOH) without germ-pore (figure d)
Trama: parallel to divergent (figure f: white arrow)
Cystidia: c.(40-50)x(12)µm, obclavate pleurocystidia (fig. g: red arrow)
Basidia: short/stubby, yellow staining in KOH with 2-4 spores (figure f)
Pileipellis: a trichoderm or palisoderm?: mycelial of non-repent partly enlarged differentiated hyphae - scalp section (figure e: black arrow)
Sample: GMA PT 3 PNA 9.7
3881 m   9°56'34.824" S 77°22'55.302" W

Observations: *Inocybe*: by odor, spore print, mycelial epidermis, dry silky umbonate cap, fibrous annulus, lack of scales, pleurocystidia, germ-pore lacking spores and terrestrial nature (see Arora (1986:455), Dennis (1970) p.71. Not a *Pholiota*, due to non-cinnamon or rust brown spore print, lack of viscid or scaly pileus, odor, nature of the annulus and non lignicolous nature.

1 unit = 3.75µm

THE MACROFUNGI OF ANDEAN PERU   Part 1

P215 *cf. **Inocybe sp.***   Spores: x̄ (10.1x5.4)µm (Q=1.9)
Northern Andes: Lambayeque

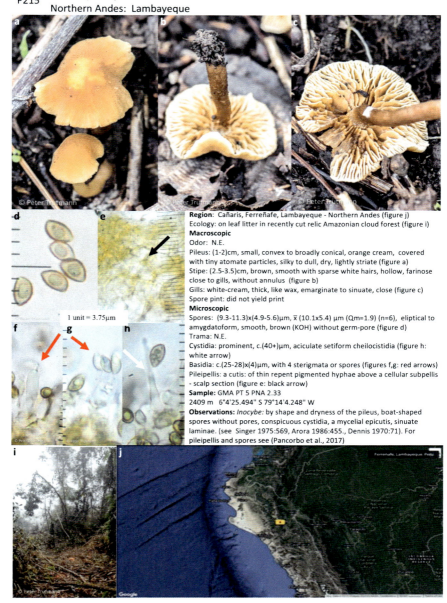

Region: Cañaris, Ferreñafe, Lambayeque - Northern Andes (figure j)
Ecology: on leaf litter in recently cut relic Amazonian cloud forest (figure i)
Macroscopic
Odor: N.E.
Pileus: (1-2)cm, small, convex to broadly conical, orange cream, covered with tiny atomate particles, silky to dull, dry, lightly striate (figure a)
Stipe: (2.5-3.5)cm, brown, smooth with sparse white hairs, hollow, farinose close to gills, without annulus (figure b)
Gills: white-cream, thick, like wax, emarginate to sinuate, close (figure c)
Spore pint: did not yield print
Microscopic
Spores: (9.3-11.3)x(4.9-5.6)µm, x̄ (10.1x5.4) µm (Qm=1.9) (n=6), eliptical to amygdatoform, smooth, brown (KOH) without germ-pore (figure d)
Trama: N.E.
Cystidia: prominent, c.(40+)µm, aciculate setiform cheilocistidia (figure h: white arrow)
Basidia: c.(25-28)x(4)µm, with 4 sterigmata or spores (figures f,g: red arrows)
Pileipellis: a cutis: of thin repent pigmented hyphae above a cellular subpellis - scalp section (figure e: black arrow)
Sample: GMA PT 5 PNA 2.33
2409 m   6°4'25.494" S 79°14'4.248" W
Observations: *Inocybe:* by shape and dryness of the pileus, boat-shaped spores without pores, conspicuous cystidia, a mycelial epicutis, sinuate laminae. (see Singer 1975:569, Arora 1986:455., Dennis 1970:71). For pileipellis and spores see (Pancorbo et al., 2017)

1 unit = 3.75µm

Family Lyophyllaceae

© Peter Trutmann

Calocybe

Calocybe sp.?   Spores: x̄ (6.0x3.5)µm (Qm=1.7)
Southern Andes: Cusco

Region: Ccorca, Cusco, Cusco - Southern Andes (figure h)
Ecology: organic material, in Ichu based puna grasslands (fig. g)
Macroscopic
Odor: agreeable
Pileus: 2cm, convex to broadly convex, white - purplish white, smooth, dull, fleshy (figures a,b)
Stipe: 3(0.5)cm, white- pinkish white, with feltlike surface covering darker base, fleshy, central, without annulus (figure a,b)
Gills: white, adnexed, crowded
Spore print: white- off white(figure b: orange arrow)
Microscopic
Spores: (5.6-6.8)x(3x3.8)µm, x̄ (6.0x3.5)µm (Qm=1.7) (n=5), subglobose, hyaline(KOH), inamyloid (Melzer's), smooth, without germore (figure c)
Trama: parallel (figure e: white arrow)
Cystidia: not found,
Basidia: with inclusions and 2 or more spores (fig. d: red arrow)
Pileipellis: a cutis: of repent, parallel hyphae scalp section (figure f: black arrow)
Sample: GMA PT 3 PSA 4.10
3991 m   13°33'55.242" S 72°1'40.314" W
Observation: *Calocybe*: white purplish basidiophore, with inclusions in basidia, regular trama, inamyloid spores, crowded gills, and spores white to cream, as well as a cutis as epidermis
See (Singer, 1975:218)

Gerhardtia

P217
(4)

Gerhardtia cibaria *(syn. Pleurocollybia cibaria)* Spores: x̄ (5.3x4.0)µm (Qm=1.3)
Southern Andes: Cusco, Apurimac

Region: Ccorca and, Ocra, Cusco, Abancay market (fig. j)
Ecology: soil associated with Ichu or below native shrubs (figures h,i)
Macroscopic
Odor: agreeable
Pileus: 2-4cm, broadly convex to plane, cream and light brown, fleshy, smooth, silky to shiny, moist to dry (figure a,d)
Stipe: 4cm, white to cream, smooth, central, smooth, without annulus (fig. b)
Gills: white to cream, emarginate-adnate, close-crowded (figure c)
Spore print: did not yield a print
Microscopic
Spores: (4.7-5.6x(3.7-4.0)µm, x̄ (5.3x4.0)µm (Qm=1.3) (n=5), oval-elliptical, smooth, hyaline with vacuoles (KOH), inamyloid (Melzer's), without germ-pore (figure d)
Trama: parallel, without clamp connections (figure f: white arrow)
Cystidia: not found
Basidia: elongated, claviform with 4 spores (white arrow) at times with inclusions (siderophores?) (figures e,f: red arrow)
Epicutis: a cutis: of enlarged often pigmented hyphae - scalp section (figure g: black arrow)
Samples:
Ccorca, Cusco: GMA PT 3 PSA 4.23 ,
4047 m,   13°33'52.596" S 72°1'29.909" W
Ocra, Anta, Cusco: GMA PT 3 PSA 5.2 ,
3762m   13°36'7.656" S 72°12'20.531" W
Market Abancay   GMA PT 1 PSA 4.1 ,
Cusco, Cusco (UNSAAC): GMA PT 2 PSA 5.11
3587m   13°34'20.586" S 71°51'20.43" W
Observations: *Gerhardtia cibaria* Matheny et al, (syn. with *Pleurocollybia cibaria* Singer Singer (1975: 274). A taxonomic change is proposed based on the presence of minutely warty basidiospores, cyanophilic bodies with siderophores in basidia and morphology like the Genus *Gerhardtia* -Family Lyophyllaceae (Matheny et al., 2017). In the limitations of this study the spores did not seem ornamented or irregular but ellipsoid-oval and smooth. Here the granular basidial content was sporadic and not tested with acetylcarmine to confirm siderophores. Known locally as 'Qoncha Cusqueña'.

REFERENCES

Alvarez Loayza, Patricia, Larry Evans and Jean D. Lodge 2014. FUNGI of Cocha Cashu. Madre de Dios, PERU: Estación Biológica Cocha Cashu, Parque Nacional MANU, Madre de Dios, PERU.

Antonelli, A., W. D. Kissling, S. G. A. Flantua, M. A. Bermúdez, A. Mulch, A. N. Muellner-Riehl and C. Hoorn 2018. Geological and climatic influences on mountain biodiversity. . Nature Geoscience 11(718-725. doi: https://doi.org/10.1038/s41561-018-0236-z

Arora, D. 1986. Mushrooms demistified. Berkeley, Ca: Ten Speed Press.

Babos, M., K. Halász, T. Zagyva, Á. Zöld-Balogh, D. Szegő and Z. Bratek 2011. Preliminary notes on dual relevance of ITS sequences and pigments in Hygrocybe taxonomy. Persoonia 26: 99–107. . doi: 10.3767/003158511X578349. PMC 3160800. PMID 22025807.

Baker, K. F. and Cook, R. J. 1974. Biological Control of Plant Pathogens. San Francisco. 433 pp.: Freeman.

Baroni, Timothy J. , Edgardo Albertó, Nicolás Niveiro and Bernardo Lechner 2012. New species and records of Pouzarella (Agaricomyetes, Entolomataceae) from northern Argentina. KURTZIANA 37: 41-63.

Birkebak, Joshua M., Jordan Mayor, Martin Ryberg and Patrick B. Matheny 2013. A systematic, morphological and ecological overview of the Clavariaceae (Agaricales). Mycologia 105: 896-906. doi: 10.3852/12-070

Boa, Eric 2004. Wild Edible Fungi a global oversview of their use and importance to people: FAO, Viale delle Terme di Caracalla, 00100 Rome, Italy

Cárdenas Medina, Anatoly , Mishari Garcia Roca, Bhushan Shrestha and Magdalena Pavlich 2019. Macrohongos de Tambopata, Madre de Dios, PERÚ. In Macrohongos de Tambopata, Madre de Dios, PERÚ, ed. Universidad Nacional de Amazonas Madre de Dios (UNAMAD). https://fieldguides.fieldmuseum.org/sites/default/files/rapid-color-guides-pdfs/1183_peru_macrofungi_of_tambopata.pdf.

Chimey Henna, C. A. and M. E. Holgado Rojas 2010 Los hongos comestibles silvestres y cultivados en Perú. In Hacia un desarrollo sostenible del sistema de producción-consumo de los hongos comestibles y medicinales en latinoamérica: avances y perspectivas en el siglo XXI ed. Curvetto N. Martínez-Carrera D., Sobal M.;Morales P., Mora, V. M., 381-395.

Co-David, D., D. Langeveld and M.E. Noordeloos 2009. Molecular phylogeny and spore evolution of Entolomataceae. Persoonia. 23: 147–176. doi: 10.3767/003158509X480944

Dennis, Richard W.G. 1970. Fungus flora of Venuzuela and adjacent countries. London: Her Majesty's Stationary Office.

Espinoza M. 2003. Hongos macroscopios de la clase basidiomycetes en el centro de investigacion Allpahuaya, Loreto-Perú. Departamento Academico de Microbiologia, , Universidad Nacional de la Amazonia Peruana.

Franco-Molano A.E., Vasco-Palacios A.M., Lopez-Quintero C.A., and Boekhout T., 2005. Macrohongos de la Región del Medio Caquetá - Colombia. Medellin, Colombia: Multimpresos Ltda., .

Franco-Molano, Ana Esperanza 1999. A new species of Macrolepiota from Colombia/ Una nueva especie de Macrolepiota de Colombia. Actual Biol 21 13-17.

Franquemont, C. , T. Plowman, E. Franquemont, S.R. King, C. Niezgoda, W. Davis and C.R. Sperling 1990. The ethonobotany of Chinchero, an Andean community in southern Peru. Fieldiana 24: 150.

García Roca, Mishari Rolando 2015. Contribución al conocimiento de los macrohongos en la provincia de Tambopata - Madre de Dios, Peru. Escuela Téchnica Superior de Ingenieros de Montes, Universidad Politechnica de Madrid.

Gazis, Romina 2007. Evaluation of the macrofungal community at Los Amigos biological station, Madre De Dio. Graduate Faculty of theColleague of Science and Engineering, Texas Christian University.

Gorzelak, Monika A. , Amanda K. Asay, Brian J. Pickles and Suzanne W. and Simard 2015. Inter-plant communication through mycorrhizal networks mediates complex adaptive behaviour in plant communities. AoB Plants. 27: plv050. doi: doi: 10.1093/aobpla/plv050

Guzman, G. Allen, J.W. Gartz, J. 2000. A worldwide geographical distribution of the neurotropic fungi, an analysis and discussion. Arch. St. Sc. Nat. 14: 189-280.

Guzman, Gastón 1997. Los nombres de los hongos y lo relacionado con ellos en América Latina. Veracruz, Mexico: Instituto de Ecología, A.C.

Hawksworth, David L. 2001. Mushrooms: The Extent of the Unexplored Potential. 3: 5. doi: 10.1615/IntJMedMushr.v3.i4.50

Herrera, Fortunato L. 1941. Sinopsis de la Flora de Cusco. In Sinopsis de la Flora de Cusco. Cusco: Universidad Nacional de San Antonio Abad de Cusco.

Hibbett D.S., R. Bauer, M. Binder, A.J. Giachini, K. Hosaka5, A. Justo, E. Larsson, K.H. Larsson, J.D. Lawrey, O. Miettinen, L.G. Nagy, R.H. Nilsson, M. Weiss and R.G. Thorn 2014. Agaricomycetes. In The Mycota vol. 7A: Systematics and Evolution Chapter: Agaricomycetes, ed. David J. McLaughlin and Joseph W. Spatafora, 373-412: Springer.

Holgado Rojas, M.E., J. Delgado Salazar, K. Pérez Leguia, N. Bautista Valverde, P. Sánchez Huamán, A. Quispe Peláez and C. Vincente Ramirez 2010. Etnomicologia en el festival del Q'oncha Raymi. Q'euña (Revista de la Sociedad Botanica del Cusco) 3: 58-59.

Holgado Rojas, M.E., M. Olivera Gonzales and J. Delgado Salazar 2007. Macomycetos de la microcuenca de Q'euña-Zurite-Anta. Q'euña (Revista de la Sociedad Botanica del Cusco) 1: 15-18.

Holguin de, Diego G. 1608. Vocabulario de la lengua general de todo el Perú llamada lengua Qquichua, o del Inca. Lima: Francisco del Canto.

Laessoe, T. and Petersen, J.H. 2019. Fungi of Temperate Europe. Vol. 1-2: Princeton University Press.

Laessoe, T. and Petersen J.H. 2008. Equatorial fungi-mycological biodiversity in Ecuador. Svampe 58: 1-27

Largent, D. L. , D. Johnson and R. Watling 1977. How to identify mushrooms to Genus III: Microscopic features Eureka, Ca: Mad River Press.

Lazo, Waldo 2001. Hongos de Chile: Atlas micólogico In Hongos de Chile: Atlas micólogico 309. Chile: Departamento de Ciencias Ecolócicas, Universidad de Chile, http://www.libros.uchile.cl/files/presses/1/monographs/424/submission/proof/files/assets/basic-html/toc.html.

Llatas-Quiroz, S. and M. López-Mesones 2005. Bosques montanos-relictos en Kañaris (Lambayeque, Perú). Relict montane forests from Kañaris (Lambayeque, Peru). Rev. peru. biol. 12: 299 - 308

Lodge, D. Jean , Mahajabeen Padamsee, P. Brandon Matheny, M. Catherine Aime, Sharon A. Cantrell, David Boertmann, Alexander Kovalenko, Alfredo Vizzini, Bryn T. M. Dentinger, Paul M. Kirk, A. Martyn Ainsworth, Jean-Marc Moncalvo, Rytas Vilgalys, Ellen Larsson, Robert Lücking, Gareth W. Griffith, Matthew E. Smith, Lorelei L. Norvell, Dennis E. Desjardin, Scott A. Redhead, Clark L. Ovrebo, Edgar B. Lickey, Enrico Ercole, Karen W. Hughes, Régis Courtecuisse, Anthony Young, Manfred Binder, Andrew M. Minnis, Daniel L. Lindner, Beatriz Ortiz-Santana, John Haight, Thomas Læssøe, Timothy J. Baroni, József Geml and Tsutomu Hattori 2014. Molecular phylogeny, morphology, pigment chemistry and ecology in Hygrophoraceae (Agaricales). Fungal Diversity 64: 1-99.

Lücking, Robert , Manuela Dal-Forno, James D. Lawrey, Frank Bungartz, María E. Holgado Rojas, Jesús E. M. Hernández, Marcelo P. Marcelli, Bibiana Moncada, Eduardo A. Morales, Matthew P. Nelsen, Elias Paz, Luis Salcedo, Adriano A. Spielmann, Karina Wilk, Susan Will-Wolf and Alba Yánez-Ayabaca 2013. Ten new species of lichenized Basidiomycota in the genera Dictyonema and Cora (Agaricales: Hygrophoraceae), with a key to all accepted genera and species in the Dictyonema clade. Phytotaxa 139(1-38.

Mata, M., Roy Halling and Gregory M. Mueller 2003. Macrohongos de Costa Rica (Costa Rica Macrofungi). Editorial INBio - Instituto Nacional de Biodiversidad (INBio), .

Mata, Milagro H. , Magdalena H. Pavlich and Teresa D. Espinoza Mori, Maribel A. 2006. Hongos de Allpahuayo-Mishana, Iquitos, Loreto, Peru. Chicago: Environmental and Conservation Programs, The Field Museum, Chicago.

Matheny, P.B., J.M. Curtis, V Hofstetter, ., M.C. Aime, J.M. Moncalvo, Z.W Ge, ., J.C. Slot, J.F. Ammirati, T.J. Baroni, N.L. Bougher, K.W. Hughes, D.J. Lodge, R.W. Kerrigan, M.T. Seidl, D.K. Aanen, M. DeNitis, G.M Daniele, ., D.E. Desjardin, B.R. Kropp, L.L. Norvell, A. Parker, E.C. Vellinga, R. Vilgalys and D.S. Hibbett 2006. Major clades of Agaricales: a multilocus phylogenetic overview. Mycologia 98: 982–995. doi: 10.3852/mycologia.98.6.982

Mohammed, J. 2019. The Role of Genetic Diversity to Enhance Ecosystem Service. American Journal of Biological and Environmental Statistics 5(3): 46-51.

Price, M. F., Ed. 2006. Global change in mountain regions. Duncow, UK, Sapiens Publishing.

Neves, Maria A, Luri Goulard Baseia, E. RIchardo Dechler-Santos and Aristóteles Góes-Neto 2013. Guide to common fungi of the semiarid region of Brazil. Florianópolis, Brazil: TECC Editora.

Palacios Noe, Lindsay K. 2015. Hongos macroscopicos del phylum basidiomycota en el bosque de neblina de cuyas ayabaca oiura. Facultad de Ciencias, Universidad Nacional de Piura.

Pancorbo, F. , M.A. Ribes, F. Esteve-Raventós, J. Hernanz, I. Olariaga, P.P. Daniëls, A. Hereza, S. Sánchez, J.F. Mateo and F. Serrano. 2017. Contribución al conocimiento de la biodiversidad fúngica del parque nacional de ordesa y monte perdido ii. Pirineos.

Pavlich Herrera, M.R. 1976. Ascomycetes y Basidiomycetes del Peru. I : con emfasis en especies de la ceja de montaña y selva tropical. Memorias del Museo de historia natural "Javier Prado" 17: 88.

Pavlich Herrera, M.R. 2001. Los Hongos Comestibles del Perú. Revista de Ciencias Biológicas BIOTA 100 3-19.

Pegler, D. N. 1965 Studies on Australasian Agaricales. Australian Journal of Botany 13: 323-356.

Pegler, P.N. and T.W.K. Young 1978. Spore form and phylogeny of Entomotaceae (Agricales). Beihefte zur Sydowia (Berger, Austria) 8: 290-303.

Price, M. F., Ed. 2006. Global change in mountain regions. Duncow, UK, Sapiens Publishing.

Quispe, M.R. Gustavo., M.L. Cruz, C.Z. Alvarado, R.P. Esquivel and Peña C.L.Z. 2006. Diversidad de macromicetas en la microcuenca de K'ayra Cusco - Peru. Cantua 13: 46-50.

Quispe Pelaez, Albino 2020. Evaluación de la diversidad de hongos alimenticios silvestres del distrito de San Jerónimo Cusco y su potencial de cultivo 2017. M.Sc. Escuela De Posgrado, Universidad Nacional de San Antonio Abad del Cusco.

Reid, Derek A and Albert Eicker 1995. The Genus Hymenagaricus Heinem. in South Africa. S. Afr. I. Bot. 61: 293- 297.

Salvador Montoya, Carlos Alberto 2009. Hongos del Parque Nac ional Yanachaga-Chemillén. In Hongos del Parque Nac ional Yanachaga-Chemillén, 18. http://www.jbmperu.org/curso/hongospnych.pdf: Center for Conservation and Sustainable Development (CCSD), Missouri Botanical Gardens.

Singer, Rolf 1963. Un hongo nuevo comestible de Sudamerica [Pleurocollybia cibaria sp. nov.]. . Bol. Soc. Argent. Bot. 10:: 207-208.

Singer, Rolf; 1975. The Agricales in modern taxonomy. Vaduz, Lichtenstein: J.Cramer.

Sourell, Susanne , D. Jean Lodge, João P.M. Araújo, Timothy J. Baroni, Priscila Chaverri, Ariadne Furtado, Tatiana Gibertoni, Fernanda Karstedt, Jadson J.S. Oliveira, Larissa Trierveiler Pereira and Julia Simon Cardoso 2018. FUNGI of Reserva Particular do Patrimônio Natural do Cristalino: Cristalino Lodge, RPPN Cristalino, Alta Floresta, Mato Grosso, BRAZIL Volume 2: . In FUNGI of Reserva Particular do Patrimônio Natural do Cristalino: Cristalino Lodge, RPPN Cristalino, Alta Floresta, Mato Grosso, BRAZIL Volume 2: , 34: UNEMAT, HERBAM, Baden-Württembergisches Brasilien-Zentrum der Universität Tübingen and The Field Museum, Chicago, USA.

Stamets P. 2005. Mycelium running: how mushrooms can help save the world. New York: Ten Speed Press, Random House.

Stamets P. 1996. Psilocybin mushrooms of the world. Berkeley, Ca: Ten Speed Press.

Trutmann, Peter and Luque, Amarilda R. 2012. Los hongos olvidados del Peru. In Los hongos olvidados del Peru. Orselina, Switzerland (also see Researchgate.net and Academia.com): Global Mountain Action.

Trutmann, Peter, Amarilda Luque, Mario López Mesones and Magdalena Pavlich 2019. Una primera guia de hongos del bosque de Cañaris - A first guide to the fungi of the cloud forest of Cañaris. In Una primera guia de hongos del bosque de Cañaris - A first guide to the fungi of the cloud forest of Cañaris. Global Moutnain Action. https://www.researchgate.net/publication/332798520_Una_primera_guia_de_hongos_del_bosque_de_Kanaris_-_A_first_guide_to_the_fungi_of_the_cloud_forest_of_Kanaris:

Usman, Muhammad, Tania Ho-Plágaro, Hannah E. R. Frank, Monica Calvo-Polanco, Isabelle Gaillard, Kevin Garcia and Sabine D. Zimmermann 2021. Mycorrhizal Symbiosis for Better Adaptation of Trees to Abiotic Stress Caused by Climate Change in Temperate and Boreal Forests. Frontiers in Forests and Global Change 4. doi: 10.3389/ffgc.2021.742392

Vizzini, Alfredo and Zai-Wei Ge 2015. Redescription of Clitocybe umbrinopurpurascens (Basidiomycota, Agaricales) and revision of Neohygrophorus and Pseudoomphalina. . Phytotaxa 219 43-57. doi: doi:10.11646/phytotaxa.219.1.3.

Wright J.E. Albertó E. 2002. Gúia de los hongos de la región pampeana. I. hongos con laminillas. Buenos Aires: Talleres Graficos Lux S.A.

COLLECTION INDEX